Flow Cytometry
Third Edition
A Practical Approach

Edited by

Michael G. Ormerod

34 Wray Park Road, Reigate,
Surrey RH2 0DE, U.K.

OXFORD
UNIVERSITY PRESS

Preface

This book is intended as a handbook for every laboratory that has a bench-top flow cytometer or a fluorescence-activated cell sorter. It is an introduction and guide to those new to the field and a first point of reference for experienced practitioners who want to investigate a new technique.

The third edition has built on the ground covered in the first and second editions. Many of the chapters have been updated. Two chapters have been dropped and three new chapters added. The chapter on immunophenotyping—the most important clinical application of flow cytometry—has been retained and strengthened by the addition of a chapter on quality control in the clinical laboratory. The utility of the book in a clinical laboratory has been further enhanced by the addition of a chapter covering 10 other clinical applications (further clinical applications). Flow cytometry has found increasing application in the field of apoptosis research. A new chapter has been added to cover this important topic.

The size of the book (and hence its cost) has been kept within reasonable limits. Every flow cytometry laboratory can afford to have a copy on the shelf as a first point of reference. The book is not fully comprehensive, but it does aim to cover over 90% of the applications of flow cytometry in mammalian biology.

In an expanding field, new developments are continually appearing. It is recommended that everyone with a serious interest in flow cytometry should join the International Society for Analytical Cytology. The membership fee includes a subscription to the journal, *Cytometry*, which gives wide cover to new developments in this and related areas. The Society's Web site (www.isac-net.org) has links to the growing number of affiliated national organizations.

A computer is an essential element of all flow cytometers. The data generated is written to computer disc in a standard format so these files can be analysed off-line by a variety of computer programs. Correct analysis of the data is essential. Examples of data files generated by the applications described in this book are available on CD-ROM (1), which serves as a companion to this volume.

Reference

1. Ormerod, M. G. (1996). *Data analysis in flow cytometry—a dynamic approach.* Published by the author on CD-ROM.

Reigate, 2000   M.G.O.

Contents

CONTENTS

CONTENTS

X

Protocol list

Abbreviations

7-AAD	7-aminoactinomycin D
ADB	1,4-diacetoxy-2,3-dicyanobenzene
ADC	analog-to-digital converter
ALG	anti-lymphocyte globulin
AO	Acridine Orange
APC	allophycocyanin
ATG	anti-thymocyte globulin
βAPP	beta-amyloid precursor protein
BCECF	2′,7′-bis-carboxyethyl-5(6)- carboxyfluorescein
BIODIPY	4,4-difluoro-4-bora-3α, 4α-diaza-s-indacene
bp	base pair
BrdUrd	5′-bromodeoxyuridine
BrdUTP	bromodeoxyuridine triphosphate
BSA	bovine serum albumin
$[Ca^{2+}]_i$	concentration of intracellular ionized calcium
CA3	chromomycin A3
CCCP	carbonyl cyanide m-chlorophenylhydrazone
CD	cluster of differentiation
CDC	Centers for Disease Control, USA
CFDA	carboxyfluorescein diacetate
CFDA SE	carboxyfluorescein diacetate, succinimidyl ester
CM-DiI	chloromethylbenzamido derivative of octadecylindocarbocyanine
CMFDA	5-chloromethylfluorescein diacetate
CMXRos	a chloromethyl derivative of X-rhosamine (MitoTracker Red)
CMTMR	5 (and 6)- ([(4-chloromethyl)benzoyl]amino) tetramethylrhodamine
CV	coefficient of variation
CVID	common variable immunodeficiency
CyA	cyclosporin A
Cy-chrome	phycoerythrin–cyanine5 conjugate
DAG	2-diacylglycerol

DAPI	4′,6-diamidino-2-phenylindole
DC	dendritic cells
DCH	2,3-dicyanohydroquinone
DCFH	2′,7′-dihydrodichlorofluorescein
DI	DNA index
$DiOC_6$	3,3′-dihexyloxacarbocyanine
$DiOC_{18}$	3,3′-dioctadecyloxacarbocyanine
DMSO	dimethylsulfoxide
DNTP	deoxynucleotide triphosphate
DOP-PCR	degenerate oligonucleotide-primed PCR
DPBS	Dulbecco's phosphate-buffered saline
dUTP	deoxyuridine triphosphate
EB	ethidium bromide
EBV	Epstein–Barr virus
ECD	phycoerythrin–Texas Red conjugate
EDTA	ethylenediaminetetraacetic acid
ELISA	enzyme-linked immunoabsorbent assay
EMA	ethidium monazide
EQAS	external quality assurance survey
FALS	forward-angle light scatter
Fc	crystallizable fragment
FCM	flow cytometry
FBS	fetal bovine serum
FDA	fluorescein diacetate
FISH	fluorescent *in-situ* hybridization
FITC	fluorescein isothiocyanate
FL1, FL2, etc.	Fluorescence parameter 1, 2, etc., on the flow cytometer
GSH	glutathione
HE	dihydroethidium
HIV	human immunodeficiency virus
HLA	human leucocyte antigen
HPA	human platelet antigen
HPC	haematopoietic progenitor cells
IdUrd	iododeoxyuridine
IFN	interferon
IgG	immunoglobulin G
IL	interleukin
Indo-1	[1-[2 amino-5-[carboxylindol-2-yl]-phenoxy]-2–2′-amino-5′-methylphenoxy] ethane $N,N,N'N'$- tetraacetic acid
IP3	inositol 1,4,5-trisphosphate
ISEL	*in-situ* end-labelling
ISHAGE	International Society for Hematotherapy and Graft Engineering
IU	International units
JC-1	5,5′,6,6′,-tetrachloro-1,1′,3,3′-tetraethylbenzimidazolylcarbocyanine iodide
K_d	effective dissociation constant

laser	light amplification by stimulated emission of radiation
LGL	large granular lymphocytes
LI	labelling index
LWP	long wavelength pass
mAb	monoclonal antibody
mBrB	monobromobimane
mClB	monochlorobimane
MDR	multi-drug resistance
MESF	molecules of equivalent soluble fluorochrome
MLR	mixed lymphocyte reaction
MMP	mitochondrial membrane potential
MRD	minimal residual disease
MRP	MDR-associated protein
MTG	MitoTracker Green FM
NEQAS	National External Quality Assurance Scheme
NK	natural killer cell
PBL	peripheral blood lymphocytes
PBS	phosphate-buffered saline
PBSA	PBS with BSA
PC5	phycoerythrin–cyanine5 conjugate
PCD	programmed cell death
PCNA	proliferating cell nuclear antigen
PCR	polymerase chain reaction
PE	phycoerythrin
PerCP	peridinin chlorophyll-A protein
PE–Cy5	phycoerythrin–cyanine5 conjugate
PE–Cy7	phycoerythrin–cyanine7 conjugate
PFA	paraformaldehyde
PHA	phytohaemagglutinin
pH_i	intracellular pH
PI	propidium iodide
PIP2	phosphatidylinsitol 4,5-bisphosphate
p.l.m.	per cent labelled mitosis
PLP	periodate/lysine/formaldehyde mixture
PMA	phorbol myristate acetate
PMT	photomultiplier
PNH	paroxysmal nocturnal haemoglobinuria
PS	phosphatidyl serine
Py	pyronin Y
QC	quality control
RALS	right-angle light scatter
RM	relative movement
RPMI	Roswell Park Memorial Institute (medium)
SBIP	strand break induction by photolysis
SD	standard deviation

ABBREVIATIONS

SNAFL	SemiNaphthoFluorescein
SNARF	SemiNaphthoRhodaFluor
SPF	S phase fraction
SV	simian virus
SWP	Short wavelength pass
T_C	cell-cycle time
Tdt	terminal deoxynucleotidyl transferase
T_{G2+M}	G_2/M transit time
TNF-α	tumour necrosis factor-α
T_{pot}	potential doubling time
T_S	S-phase transit time
TUNEL	Tdt-mediated dUTP nick end-labelling

Chapter 1

Introduction to the principles of flow cytometry

Nigel P. Carter* and Michael G. Ormerod[†]

*Sanger Institute, Hinxton Hall, Hinxton, Cambridge CB10 1RQ,UK

[†] 34, Wray Park Rd, Reigate, Surrey RH2 0DE, UK

1 Introduction

Flow cytometry is a technique for making rapid measurements on particles or cells as they flow in a fluid stream one by one through a sensing point. The important feature of flow cytometric analysis is that measurements are made separately on each particle within the suspension in turn, and not just as average values for the whole population. The ability of laser- and arc lamp-based flow cytometers to measure multiple cellular parameters, based on light scatter and fluorescence, and to purify physically subpopulations of cells has led to the increasingly widespread use of this instrumentation in biology and medicine. The applications of flow cytometry and cell sorting are numerous. A wide range of fluorescent probes is available for directly estimating cellular parameters such as nucleic acid content, enzyme activity, calcium flux, membrane potential, and pH. Conjugation of fluorescent dyes to ligands and to polyclonal and monoclonal antibodies has enabled the density and distribution of cell-surface and cytoplasmic determinants and receptors to be studied, as well as allowing functional subpopulations of cells to be identified. Many of these fluorescent dyes and reagents can be used in combination to produce multiple correlated measurements. For example, it is now commonplace in immunology to measure two light-scatter and three immunofluorescence parameters on each cell.

The use of flow cytometers for cell sorting is also widespread. Applications range from the separation of large numbers of cells for functional studies or chromosomes for preparing gene libraries to the direct cloning of single, rare, transfected or hybridoma cells into each well of a tissue culture plate.

Unfortunately, flow cytometers are not simple instruments. As with all sophisticated measuring devices, it is important to possess a basic knowledge of the underlying principles of operation so that the significance and accuracy of the results can be assessed. For example, the quality of sample preparation and staining is as important in the precision of measurements as the design of the fluidic, optical, and electronic components of the instrument itself.

Within the scope of this introduction it is not possible to explain the theory of operation and design of flow cytometers in detail. However, we will try and explain in relatively simple terms the basic principles of flow cytometric practice. More detailed information can be found in refs 1–4.

2 Techniques for sample preparation (see also Chapter 3)

The aim of sample preparation is to produce a suspension of single particles, stained in a specific way, which will pass through the system without disrupting the smooth flow of fluid or blocking tubes or orifices. The particles analysed may be whole cells, cell organelles, or specific clumps of tissue such as Islets of Langerhans.

Producing a suspension of individual particles from biological samples can range from being straightforward to frustratingly difficult. Body fluids, in particular blood, generally contain individual cells that can be stained and processed directly on the flow cytometer. For example, a sample of blood can be stained with Thiazole Orange orange and analysed on the flow cytometer to obtain a count of reticulocytes (see Chapter 7, Section 2). Sometimes, the sample is enriched before analysis if the cell of interest is relatively rare. For example, blood leucocytes constitute less than 0.1% of all the cells in peripheral blood and it is customary to lyse the red blood cells to obtain a sample of white blood cells for analysis.

Solid tissues present a much more difficult and varied problem for flow cytometry. A technique which successfully releases cells from one tissue can fail totally on another. For example, lymphoid tissues can usually be prepared by simple chopping and teasing of the organ followed by sieving and density-gradient centrifugation to provide a sample of separated lymphocytes. Other organs and solid tumours require enzymatic digestion before a good yield of individual cells is obtained. Detergents are often required in the preparation of organelles, such as isolated nuclei and chromosomes, where it is necessary to remove cell membranes and cytoplasm.

With all preparative methods great care must be taken to ensure that the technique itself does not bias the results. For example, density-gradient centrifugation of lymphocytes may preferentially enrich some subpopulations. Enzymatic preparative techniques can alter cell-surface antigens and affect cell viability.

Usually, cells are stained by incubation, under appropriate conditions, with a fluorescent dye or fluorescent-conjugated antibody or ligand. For accurate interpretation of results, it is important that the staining is specific for and proportional to the feature to be measured. Unfortunately, it is not unusual for fluorescent probes and even monoclonal antibodies to bind non-specifically and care must be taken to block cross-reactions. Heterogeneity in the uptake of dyes by cells also results in degraded resolution of the measured parameter, unless the heterogeneity itself is of interest.

Optical crosstalk can occur when combinations of fluorochromes with overlapping emission spectra are used together. The result of this spectral overlap is that a proportion of the fluorescence from one dye reaches the detector intended to measure the fluorescence of the second dye, and vice versa. While this effect can be easily corrected using optical filtration or electronic compensation, the inevitable compromises introduced can reduce the overall sensitivity of measurements. Energy transfer between fluorochromes in close proximity can also take place (see Chapter 2, Section 2). Here the energy absorbed by one dye is transferred to the second dye, which then fluoresces. The consequence is that fluorescence from the first dye is quenched, producing an underestimate of fluorescence intensity, while the second dye is inappropriately excited, producing an overestimate of its intensity. However, this phenomenon can be directly exploited to measure the proximity of structures in or on cells.

The problems of sample preparation and staining are easily assessed by the use of appropriate controls designed to measure the specificity and accuracy of the measurements.

3 Fluidics

The fluidic system of a flow cytometer is used to deliver particles of a random three-dimensional suspension singly to a specific point in space intersected by the illuminating beam. This is generally achieved by injecting the sample suspension into the centre of an enclosed channel through which liquid is flowing. Channels have been designed so that the tightly focused core of sample remains separated from the surrounding sheath of fluid. Typical flow chambers can transport particles to the detection point with an accuracy of better than ± 1 micron.

3.1 Principles of fluid flow and hydrodynamic focusing

There are two principles of fluid flow used in the design of flow chambers: (a) laminar flow with viscous drag; and (b) turbulent boundary drag.

(a) In laminar flow, the velocity of the fluid front at the inlet is the same both at the walls and in the centre of the channel. As the fluid moves away from the inlet down the channel, viscous drag at the walls slows the outer layers of liquid. The fluid front changes into a parabola with the greatest velocity at the centre of the flow channel and zero velocity at the flow channel wall (see *Figure 1a*). This velocity gradient draws particles towards the centre in a process known as hydrodynamic focusing. Further down the flow channel the velocity front forms a stable parabola so that the particles continue to flow at the centre of the channel. The distance from the inlet to the formation of the stable velocity parabola is known as the inlet length and is approximately 50 times the channel diameter.

(b) In turbulent boundary flow, the sample is injected into a chamber rapidly tapering to a small exit orifice. The speed of entry of the sample combined

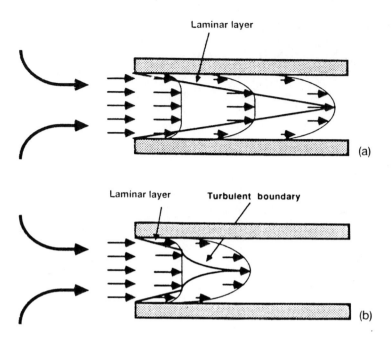

Figure 1 The principles of fluid flow used in flow chamber design. (a) Laminar flow with viscous drag; (b) turbulent boundary drag.

with the back pressure generated by the exit orifice results in sample turbulence. The turbulence at the interface between the sample and sheath produces a viscous boundary. The drag generated at this boundary forms a velocity parabola within a shorter distance than for laminar flow (see *Figure 1b*). Particles are again drawn towards and contained within the centre of the flow.

3.2 The fluidic system

A diagram of a typical fluidic system is shown in *Figure 2*. The fluidic system consists of two fluid lines feeding the flow chamber, the sheath fluid line, and the sample line. In normal operation the sheath fluid flows continuously and is controlled by regulated positive air pressure acting on the sheath reservoir. Sample flow rates are controlled by a second pressure regulator acting on the sample chamber. Differential or individual pressure gauges are used to set optimum flow conditions. An alternative approach, particularly for the accurate control of sample flow, is the use of a finely controlled syringe pump. A purge line is also often connected to the sheath inlet to allow a vacuum to be applied for clearing blockages and air bubbles.

At least one system (the Partec PAS) uses negative pressure (a vacuum) applied to the exhaust lines rather than positive pressure applied to the sheath and sample inlets to generate the fluid flow.

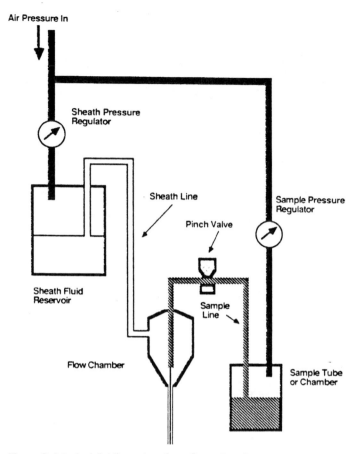

Figure 2 A typical fluidic system for a flow cytometer.

3.3 Flow chamber design

The general design of flow chambers can be divided into analytical and sorting chambers and into the use of laminar flow or turbulent boundary conditions. In general, flow chambers are designed with a wide inlet for sheath fluid, which tapers to either an exit orifice or a cylindrical or square-sided channel. The smaller bore tube used for sample injection is located at the centre of the wide inlet. In laminar flow designs, the tapering flow reduces the diameter of the core of sample before entry into the flow channel where the final hydro-dynamic focusing of the sample takes place. In turbulent boundary designs, the tapering design itself brings about hydrodynamic focusing within a short distance of the inlet. Some typical chamber designs, as used in commercial instruments, are shown in *Figure 3*.

3.3.1 Analytical chambers

These chambers generally use laminar flow conditions in their design and fall into two groups. The first group of chambers is used in laser-based flow systems

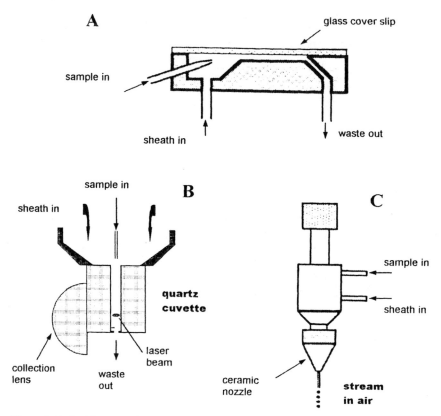

Figure 3 Typical flow chamber designs. (A) PAS II flow chamber (microscope-based). (B) Quartz cuvette (Beckman Coulter). (C) FACStar flow chamber (Becton Dickinson).

where fluorescence is measured at right angles to the illuminating beam. Most systems use a flat-sided cuvette designed to minimize unwanted light reflections and are usually positioned vertically to the laser beam. In analytical systems, the flow of fluid is usually upwards to allow air bubbles to be easily removed from the chamber. Scatter measurements can be made both at right angles to the excitation beam and in the forward direction. However, in the forward direction it is necessary to use a blocker bar to eliminate unscattered laser light.

The second group of chambers is used in microscope-based flow systems where fluorescence is measured in line with the optical path. Chamber design is constrained by the limitations of this optical system, with the chamber re-placing the horizontal microscope stage (e.g. see *Figure 3a*). The top surface of these chambers is usually a glass coverslip, so that immersion objectives of high numerical aperture can be used. Scatter measurements are restricted to within the direct optical path and are generally difficult to obtain. Some systems do not use an enclosed channel but simply squirt the hydrodynamically focused sample at a low angle across a microscope slide to be aspirated by a vacuum waste line.

Analytical chambers typically use a minimum channel bore of 250 μm to help prevent blockages and unwanted reflection from the walls. Unrestricted, the flow through such a channel would require large volumes of sheath fluid and would be difficult to control. To reduce the flow and to obtain control over operating pressures, a restriction in the exhaust from the chamber is used. Many designs use a coiled length of narrow-bore tubing to provide this resistance to the flow.

3.3.2 Flow chambers for cell sorters

The most common principle used for cell sorting is the electrostatic charging of droplets as applied to laser-based flow systems. The flow chamber design incorporates an exit orifice, usually a watchmaker's jewel or a precision drilled hole, which produces a jet of fluid. Stable droplet formation takes place under the influence of an applied oscillation. The exit orifice also provides the required resistance to control sheath flow rates.

There are two basic designs of droplet sorting chambers in general use. The first type (see *Figure 3b*) uses laminar flow conditions in a square channel cuvette where cells intersect the laser beam above the exit orifice. The second type (see *Figure 3c*) uses turbulent boundary conditions, and cells intersect the laser beam immediately below the exit orifice in the exhaust jet. This second type of analysis and sorting design is known as 'stream in air' or 'jet in air'.

Scatter and fluorescence measurements are made in the same way as analytical, laser-based, flow systems with the exception that a second blocker bar is required for 'stream in air' analysis and sorting. This blocker bar is placed in front of the right-angle collection lens to block laser light reflected by the cylindrical surfaces of the jet.

Microscope-based flow cytometers are less well suited for cell sorting. However, Partec PAS systems achieve sorting by the use of a piezoelectric fluidic valve operating on one arm of a 'Y'-shaped flow channel. This principle does not have the particle size restrictions imposed by jet and droplet formation and has been used for sorting large particles such as whole Islets of Langerhans. Another approach has been adopted in the FACSort and FACSCalibur instruments. Here a small collector arm is moved into the sample stream to selectively intercept cells of choice (see Chapter 4, *Figure 3*).

4 Detection and measurement

The flow cytometers, which are the subject of this book, make measurements based on light as the source of excitation. Intense illumination is required because cells are small and pass through the detection point rapidly. In addition, the light source must be capable of producing specific wavelengths that can be used to excite fluorescent dyes. The scattered and fluorescent light generated by cells passing through the illuminating beam is collected by photodetectors which convert the photon pulses into electronic signals. Further electronic and computational processing results in graphic display and statistical analysis of the

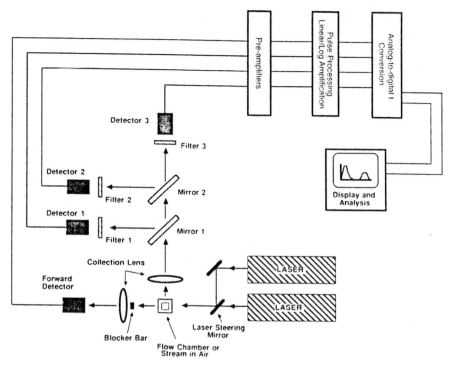

Figure 4 A generalized flow cytometer system.

measurements being made. A generalized flow cytometer system is shown in *Figure 4*.

4.1 Illumination and beam shaping

The two types of light source commonly used in flow cytometers are arc lamps and lasers. An arc lamp is a glass envelope containing a gas or vapour at high pressure. An initial, high-voltage spark between two electrodes within the envelope forms a luminous plasma arc which is maintained by the application of a high current at low voltage. The arc is intrinsically unstable and is prone to flicker and wander. Flow cytometers using epi-illumination of the stream are able to use the spot of highest intensity formed at the cathode as the focal point for the lamp condenser lens. Even illumination is then obtained by using Köhler optics where the lamp condenser lens acts as a uniformly illuminated disc. (For a description of Köhler illumination, see *Light microscopy in biology: a practical approach*, ed. A. J. Lacey, this series.) The average life of arc lamps is short and is at best only a few hundred hours.

The name laser is an acronym for 'light amplification by stimulated emission of radiation'. The laser produces a coherent, plane-polarized, intense, narrow beam of light at specific selectable wavelengths. A typical gas laser is shown schematically in *Figure 5*. Mirrors, positioned at each end of the resonator, form an optical cavity within which the plasma tube is located. The plasma tube contains a gas at a critical pressure which fluoresces under the application of a current, emitting light in all directions. The applied current raises electrons of the gas atoms into higher energy orbits. Light is produced when these electrons in high-energy orbits spontaneously decay to the ground state. The wavelength of the released photon is dependent upon the energy difference produced by the transition of the electron from the high- to the low-energy orbit.

Light emitted from the ends of the plasma tube is reflected by the mirrors back into and along the tube. When these reflected photons strike an atom in an excited state, a second photon is produced, which is of the same wavelength and phase as the stimulating photon and travels in the same direction along the plasma tube. These photons stimulate further photon release in a chain re-action that produces light amplification. The magnitude of the amplification increases with the length of the plasma tube, and is controlled by the rate at which electrons are pumped to higher energy orbits by the applied current. Flat, orthogonal windows at the ends of the plasma tube would reflect too much light to enable lasing to take place. Therefore, the windows are cut to Brewster's angle at which reflection is at a minimum. A consequence of using these Brewster windows is that the laser beam is plane-polarized.

Between 95% and 99% of the light produced within the plasma tube is re-quired to maintain lasing, so that only between 1% and 5% of the laser light can be used. The front mirror is designed to transmit the appropriate percentage of light to form the usable laser beam.

The light produced by the laser is a mixture of the specific wavelengths (laser lines) defined by the finite energy levels that the electron orbits can achieve. Selection of specific lines is produced by the use of mirror coatings which efficiently reflect only the desired wavelength. Alternatively, a prism can be placed in front of the high reflector (see *Figure 5*) to separate physically the laser emission. Light dispersed off the optical axis by the prism is not reflected back into the plasma tube and cannot stimulate lasing. Rotating the prism will alter the wavelength that will lie on the optical path, so enabling one from the range of possible lasing lines to be selected.

The lasers used in flow cytometers are atomic (e.g. helium–neon), ionic (e.g. argon-ion ion or krypton), molecular (e.g. helium–cadmium), or liquid (e.g. dye

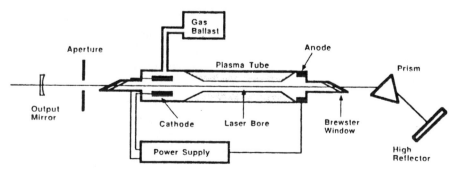

Figure 5 A typical gas laser.

lasers). Solid-state lasers are becoming available in a form suitable for flow cytometry and have the potential to be the laser of choice in the future.

The ideal laser beam shape would be a line orthogonal to and wider than the flow core, but whose thickness is less than the diameter of the smallest particle to be measured. This beam shape would maximize optical resolution of the particle passing through the spot while minimizing the positional sensitivity of the particle trajectory. In practice, a beam of this ideal shape is difficult to achieve and is usually approximated by the use of lenses producing ellipsoid or simple circular beams. Ellipsoid beams are produced by the use of crossed, cylindrical, paired lenses or by the use of spheral–cylindrical lens combinations. For example, one of the crossed, cylindrical, paired lenses used in the Epics Elite (Beckman Coulter) produces an ellipsoid focal spot of approximately 6 by 80 microns from a laser beam of 0.9 mm diameter and a wavelength of 488 nm.

'Stream in air' cytometers provide additional optical design considerations, as the stream itself acts as a half cylindrical lens in the illumination of the particle. In addition, the illuminating beam must be of a low numerical aperture to minimize the spheral aberration of the stream acting as a lens in the light path of the image. A beam wider than the stream would also produce unacceptable noise at the scatter detectors due to the stream perturbations introduced by the use of the transducer during sorting (see Chapter 4). 'Stream-in-air' cytometers use crossed, cylindrical or spheral–cylindrical lenses producing spot sizes of typically 20×60 microns. Alternatively, simple long focal length lenses producing circular spots of typically 30–50 microns in diameter are used to simplify the optical design at the expense of resolution.

4.2 Collection optics

For routine analysis, the forward collection lens gathers scattered light from approximately 1 to 20 degrees off the laser beam axis. The exact angle depends on the geometry of the system and is different for different systems. Some systems allow the user to change the angle of light collection. This angle of light minimizes the effect of refractive index changes on forward scatter measurements, so maximizing the dependence on particle size. A lens of a numerical aperture of at least 0.3 is required to collect this cone of light, a specification achieved by a simple, long working distance lens.

For the greatest sensitivity in the measurement of fluorescence, the right-angle lens is designed with a high numerical aperture to collect light over the greatest cone possible. However, as the numerical aperture increases, the working distance of the lens (the distance from the front surface of the lens to the object plane) decreases. The physical dimensions of flow chambers limit the minimum working distance and so the maximum numerical aperture which can be used. The design of most commercial sorting flow chambers limits the numerical aperture of the collection lens to 0.6 or less. However, flow cytometers using non-sorting glass cuvettes or coverslip flow chambers are able to use lenses with numerical apertures of 1 or more. The light-gathering efficiency

with these chambers can be increased further by the use of immersion objectives, which eliminate light loss due to refraction at the glass–air interface. In
some designs of cell sorter, which use a cuvette flow chamber, a small lens is
glued to the side of the cuvette, again to maximize light collection (see
Figure 3b).

4.3 Optical filtration

4.3.1 Introduction

In cytometers employing arc lamps, filters are used to select the correct wavelength of exciting light. However, most instruments use lasers which give
monochromatic light so that further filtration is unnecessary. On the output
side, filters are needed to separate the mixture of scattered and fluorescent
light collected from stained particles so that specific independent but correlated
measurements can be made. The separation of different wavelengths is
achieved by the use of dichroic mirrors and interference and absorption filters.

There are two types of glass filters in use: coloured glass and interference
filters. The details of their design can be found in the manufacturers' cata-
logues, which also give representative graphs of the spectral properties of their
filters. An individual spectrophotometer trace should also be supplied with each
filter.

Chapter 2 gives information about the excitation and emission wavelengths
of different fluorophores, and this should be used as a guide to the selection of
sets of filters. An example of a typical arrangement is given in Section 4.4
below.

A useful discussion on the selection and use of optical filters together with
further references will be found in ref. 5.

4.3.2 Coloured glass filters

The most useful of these are the long-pass filters which transmit light above a
given wavelength and absorb light of lower wavelengths. They can be obtained
with a cut-off between 300 and 700 nm. The amount of light absorbed will
increase if the filter is tilted since it presents a greater thickness to the light
beam but, apart from this, the optical properties are unchanged. Some coloured
glass filters fluoresce slightly under UV light.

Bandpass filters in coloured glass seldom have sufficient discrimination to be
of use in flow cytometry.

4.3.3 Interference filters

These consist of dielectric layers deposited *in vacuo* on a glass substrate. Depending on the thicknesses of the layers and the wavelength of the light, the
internally reflected beams interfere with one another either destructively or
constructively allowing some wavelengths to pass through the filter while
others are reflected.

When an interference filter is viewed, one surface has a metallic appearance;

this should be placed towards the light source. In this way the amount of heat absorbed in the filter is minimized (this is particularly important when a filter is placed in the primary light beam) and any fluorescence from an associated coloured glass filter is also minimized.

Interference filters can deteriorate with time and their properties should occasionally be checked in a spectrophotometer to ensure that they perform to their original specification. Spectral properties should always be measured at the orientation of use since they depend on the angle of the filter in the light beam; tilting a filter shifts its spectral properties to shorter wavelengths.

Interference filters come in two forms—bandpass and edge filters.

Bandpass filters transmit light of the desired wavelength over a narrow band (see *Figure 6*). In their simplest form they consist of two reflecting layers separated by a dielectric layer of exactly one half-wavelength thickness. This is referred to as a cavity—commercial filters generally have from one to four cavities. The basic interference filter is often combined with a blocking filter to cut out transmission from sidebands.

The major parameters of bandpass filters are the peak wavelength of transmission, the percentage of light transmitted at the peak wavelength (peak transmission), the bandwidth (usually measured as the separation in nanometres of the 50% transmission points), and the maximum transmission outside the bandpass. The shape of the transmission curve is also important and this can be estimated by comparing the bandwidths at 50% and 10% transmissions. The greater the number of cavities, the sharper the cut-off at either side of the central band.

Edge filters are either short wavelength pass (SWP) filters, which transmit light below a given wavelength and reflect light of a longer wavelength, or long wavelength pass (LWP) filters, which work in the reverse way (*Figure 7*). They are generally used in the flow cytometer as dichroic mirrors (often called beam splitters) at an angle of 45° to the light beam. For the reason given above, their spectral properties should always be measured at this angle.

The major parameters of edge filters are the cut-off (for LWP) or the cut-on (for SWP) wavelength (the wavelength for 50% transmission), the peak trans-

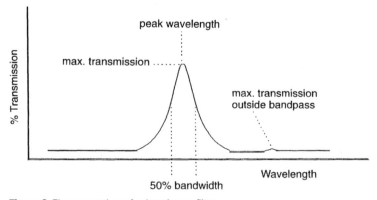

Figure 6 The properties of a bandpass filter.

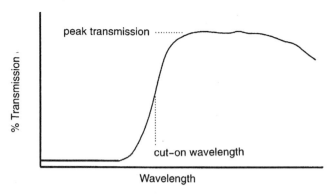

Figure 7 The properties of an edge filter.

mission, and the slope. The latter parameter defines sharpness of the cut-off and is measured between the wavelengths for 80% and 5% transmission.

The cut-off for interference filters is far sharper than that found with coloured glass.

4.3.4 Optical configuration

A typical optical configuration that could be used for the measurement of scattered light, fluorescein (FITC—green) and phycoerythrin (PE—orange) immunofluorescence is shown in *Figure 8*. Dichroic mirrors are placed at 45° to the incident beam while the absorption and interference filters are orthogonal. The first mirror in the optical path at right angles to the laser beam is a 500 nm long-pass dichroic filter, which reflects wavelengths shorter than 500 nm (the 488 nm scattered laser light) towards the right-angle scatter detector. Longer wavelengths pass on towards the second mirror, a 560 nm short-pass dichroic filter. Wavelengths greater than 560 nm are reflected towards the PE fluorescence detector and through a filter centred at 578 nm with a 28 nm half-peak bandpass. The shorter wavelengths between 500 and 560 nm incident at the second mirror pass on towards the FITC fluorescence detector and through a 530 nm filter with a 30 nm half-peak bandpass. More colours could be measured by introducing further mirror, filters, and detectors.

The emission spectra of FITC and PE and the spectral characteristics of the two bandpass filters are shown in *Figures 9* and *10*. It can be seen from *Figure 9* that, although the bandpass filters select separate wavebands, the emissions from the two fluorochromes overlap such that some fluorescence from one fluorochrome will pass to the detector intended to measure the fluorescence from the other and vice versa. This spectral overlap can be corrected during signal processing (see Section 4.5). Most modern instruments can measure at least four colours. The problem of spectral overlap then becomes more acute.

4.4 Detection devices

The two types of devices used in flow cytometers to detect scatter and fluorescence are PIN diodes and photomultiplier tubes (PMTs). PIN diodes are cheap

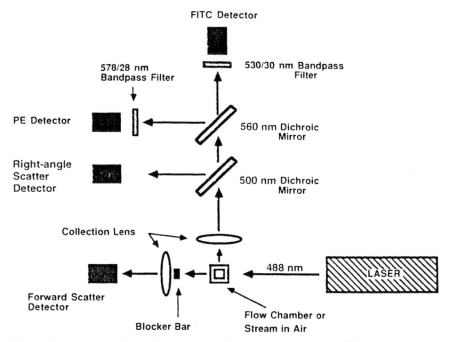

Figure 8 An optical configuration for the simultaneous measurement of forward scatter, right-angle scatter, fluorescein (FITC) and phycoerythrin (PE) fluorescence.

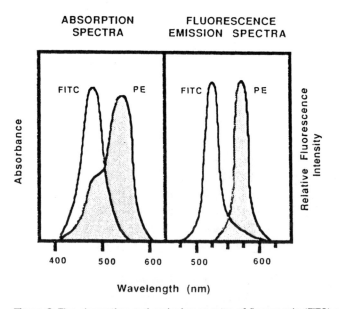

Figure 9 The absorption and emission spectra of fluorescein (FITC) and phycoerythrin (PE).

14

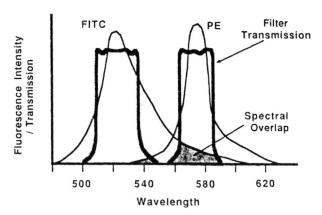

Figure 10 Filter transmission characteristics for the separation of fluorescein (FITC) and phycoerythrin (PE) fluorescence.

solid-state detectors of relatively low sensitivity but wide spectral characteristics and fast response. The lack of sensitivity of PIN diodes limits their usefulness and they are generally restricted to the measurement of forward scattered light. Photomultipliers are photosensitive electron tubes with a more restricted spectral response, but with high gain and good signal-to-noise characteristics suitable for the detection of weak fluorescence. The spectral sensitivity of PMTs is determined by the composition of the light-sensitive photocathode and this is an important factor in determining the sensitivity of a flow cytometer.

4.5 Signal processing

Light falling on the photodetector surface generates a current that is fed into a filtering pre-amplifier. The output of the amplifiers is a smoothed voltage pulse usually of between 0 and 10 volts. The amplitude of this pulse is proportional to the number of photons reaching the photodetector. Pulse shape is determined by the size and speed of the particle, the width of the illuminating beam, and, in the case of fluorescence, the distribution of the fluorochrome within the particle. The measurement of small particles using narrow illuminating beams or in 'stream in air' flow chambers will produce pulses with very fast rise times. The pre-amplifiers are designed to follow both these rapid pulses as well as the slower pulses from larger particles without distortion.

The output from the pre-amplifier will inevitably contain some background noise. It is not desirable for further signal-processing circuits to process this noise, which may be at a high frequency and could mask true pulses. This is prevented by the use of a system threshold so that further processing only takes place when the input voltage rises above a pre-set value. Usually the threshold is set on a single signal that triggers all other measuring circuits. Forward scatter is often used as the system trigger pulse, particularly when fluorescence measurements are being made. If fluorescence was used as the trigger, negative cells might not produce signals above threshold and so would not be detected or measured. An exception is in the measurement of DNA, in which case the DNA–dye fluorescence is used as the trigger since only particles containing DNA are of interest. The thresholding parameter is often referred to as the discriminator.

The pulses' output by the pre-amplifiers are generally too fast to allow measurement and display circuits to function properly. Therefore, the pulse processing circuitry provides an analog voltage memory of the signal which is held long enough for further processing to take place. The sustained voltage output from this 'sample and hold' circuit can be derived from the input signal in three ways—the output voltage can be made proportional to the height (peak height), the area (integral), or the width (time of flight) of the pulse.

Sample-and-hold circuits maintain their voltage output for a fixed period, usually between 15 and 120 μsec, and are triggered when the signal voltage rises above the pre-set threshold. During the period while the sample-and-hold circuit is operating, the electronics are unavailable for the processing of other pulses. This period is known as the 'system dead time' and events occurring during this processing period pass undetected. A large system dead time has less consequence for analysis than for sorting. The cells undetected during the system dead time will be taken randomly from the population of cells being analysed and no bias of the final results will take place. However, the time taken to acquire a specific number of events will be increased. A large system dead time during sorting will not only reduce the maximum sorting rate but will also increase the chance of undetected cells being sorted coincidentally, thus compromising sorting purity.

The various pulse-processing modes enable different measurements to be made from the same signal. Pulse width is related to the size of the particle or area of fluorochrome staining. The width of the pulse is composed of the width of the illuminating beam plus the width of the particle (and minus the small triggering and de-triggering widths). The constant offset due to the beam width can be electronically removed to produce a measurement directly proportional to the particle size.

Generally, a linear relationship between the measured fluorescence intensity and the number of fluorochrome molecules is required. If the width of the illuminating beam is less than the particle width, only a fraction of the fluorochrome molecules will be excited at one time and pulse peak measurements will only reflect the highest fluorochrome concentration and not the total content. In this case, pulse-area processing will integrate the fluorescence as the particle passes through the narrow illuminating beam so that the output becomes proportional to the total dye content. The difference between either pulse peak and area or pulse width and area can be utilized to discriminate doublet particles as applied in cell-cycle analysis (see Chapter 6).

Signals can be processed directly or after passage through a logarithmic amplifier. The relationship between the input and output voltages of a four-decade logarithmic amplifier is shown in *Figure 11*. An input voltage of 10 millivolts produces a 2.5 volt output, while a 10-fold increase in the input voltage to

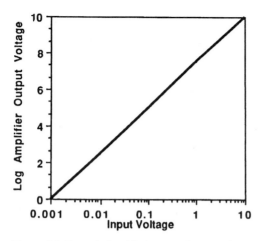

Figure 11 The relationship between input and output voltages for a 4-decade logarithmic amplifier.

100 millivolts only doubles the output to 5 volts. In this way logarithmic amplification amplifies weak signals and compresses large signals. Logarithmic amplification not only increases the resolution of weak signals but also increases the dynamic range of the measurements so that both weak and strong signals can be displayed on the same scale. The use of 10-bit analog-to-digital converters with 1024-channel resolution is now universal. With linear amplification, a 1000-fold difference in signal intensity can be displayed. However, the use of a four-decade logarithmic amplifier allows signals with a 10 000-fold difference to be displayed on the 1024-channel scale. An example of the same sample analysed using both linear and logarithmic amplification is shown in *Figure 12*.

Spectral overlap of the emission from two fluorochromes can also be corrected during signal processing. The spectral overlap produces a small signal from one fluorochrome at the detector intended to measure the fluorescence from the second fluorochrome, and vice versa. Simple circuitry can be used to electronically subtract the proportion of the fluorescence due to spectral overlap from each pulse. Examples of two-colour fluorescence measurements before and after spectral overlap correction are shown in Chapter 5, *Figure 3*.

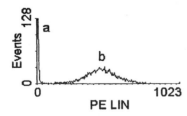
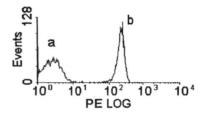

Figure 12 Linear and logarithmic displays of the immunofluorescence of a phycoerythrin-labelled (PE) monoclonal antibody to the CD4 antigen. Panel A, linear; panel B, logarithmic amplification (a, negative cells; b, CD4+ cells).

17

4.6 Digital conversion

For analysis and display by computer systems or pulse-height analysers the held voltages from the analog circuitry are digitized. The analog-to-digital converter (ADC) translates the continuous voltage analog range into a discrete scale which can be represented by a binary number. The resolution of the measurements is dependent upon the scale interval of the conversion. ADCs providing 10-bit resolution will divide the scale into 1024 elements (10 mV per division) with voltages being represented by a binary number between 0 and 1023.

4.7 Analysis and display

The result of the voltage pulses is a stream of numbers which need to be processed by a computer to display meaningful data. The most common and useful forms of display are the frequency histogram and the dual-parameter correlated plot, often known as a cytogram or dot plot. The frequency histogram is a direct graphical representation of the number of events occurring for each channel of the ADC (i.e. counts against intensity). The cytogram or dot plot is a two-dimensional extension of the frequency histogram. In this case, the locations in memory correspond to a two-dimensional array of the channels of one ADC correlated against the channels of a second. Each location within the array is incremented according to the digitized values produced by the two ADCs. The memory can then be read on to the screen to produce a square plot where each cell is represented at the co-ordinates appropriate to the measured values.

A typical frequency histogram is shown in *Figure 12*. A typical cytogram is shown in *Figure 13*. This display shows the correlated measurements of forward light scatter plotted against right-angle light scatter for a sample of peripheral blood leucocytes. Here, three main clusters can be seen which relate to the three major cell types, namely: lymphocytes, monocytes, and polymorphonuclear granulocytes.

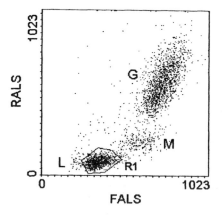

Figure 13 A cytogram or dot plot. A sample of human peripheral blood leucocytes. The intensity of right-angle light scatter (RALS) is plotted against the intensity of forward-angle light scatter (FALS) for each cell. L, lymphocyte cluster; M, monocyte cluster; G, granulocyte cluster; R1, region of interest encircling the lymphocyte cluster.

Statistical analysis of these displays is a straightforward task for the computer. Markers can be set at specific channels of a histogram and the percentage of total, mean value, and many other statistics generated for these selected events. Cytograms are usually analysed by setting boxes or polygons around areas of interest (called regions).

More complex analyses use the ability of the program to set gates or windows on defining parameters. Only cells which fall within the gates are analysed further. An example is shown in *Figure 14*. Peripheral blood leucocytes were stained with a monoclonal antibody specific for CD4-positive lymphocytes. The immunofluorescence profile generated without gating is shown in *Figure 14A*. Ungated, 10.7% of the cells display positive staining. However, if a gate is placed around the lymphocyte cluster on the forward- against right-angle scatter cytogram (region of interest 1, *Figure 13*) the profile shown in *Figure 14B* is generated; in this case 39.0% of the gated cells are positive. Both of these results are valid but represent different analyses of the same sample. The ungated data generates the number of CD4-positive cells expressed as a percentage of all leucocytes (lymphocytes, monocytes, and granulocytes). Gating enables the number of CD4-positive cells to be expressed as a percentage of lymphocytes only. Gating techniques are very powerful and allow sophisticated questions about cell and organelle subpopulations to be studied.

The histogram and the cytogram are the most used form of data display and can be updated by the computer in real time while a sample is being analysed. For publications, the data can be presented in other forms such as the contour plot, density plot, and the isometric plot. The contour plot is a cytogram where elements of similar event frequency are joined by contours (*Figure 15A*) while, in the density plot, a similar effect is achieved using different colours or grey

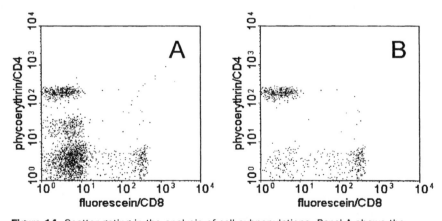

Figure 14 Scatter gating in the analysis of cell subpopulations. Panel A shows the immunofluorescence profile of the human peripheral blood leucocytes (see *Figure 13*) stained with fluorescein-labelled anti-CD8 and phycoerythrin-labelled anti-CD4 monoclonal antibodies. All cells, including lymphocytes, monocytes, and granulocytes, contribute to this profile. Panel B shows the profile obtained using the gate displayed as region of interest R1 in *Figure 13*. Only cells with scatter characteristics falling within this gate (i.e. the lymphocytes) contribute to the profile displayed in Panel B.

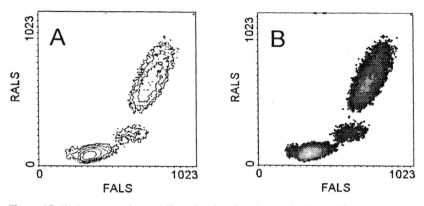

Figure 15 (A) A contour plot and (B) a density plot (data as in *Figure 13*).

scales (*Figure 15B*). The isometric plot uses the Z axis to display the frequency of correlated events and has the attractive feature of variable rotational and tilt viewing angles (see *Figure 16*).

Data can be processed immediately into histograms and cytograms (real-time analysis) or the individual correlated measurements stored on disc in time sequence (list mode). List-mode data has the advantage that the data may be re-processed, gated, and displayed as often as required. This feature is very useful when it is not clear where to set the gates and a more considered analysis is required, although real-time analysis is the most efficient approach for rapid results. Flow cytometers generate vast amounts of data. Analysing cells at 5000 per second and measuring six parameters on each cell will eventually fill up a hard disc. It is unusual to find researchers willing to delete data, valuable or not, so it is important to have as great a capacity for data storage as is possible. High-capacity tape drives, optical discs, or writable CD-ROMS are ideal for removable storage. It should also be remembered that magnetic media is not permanent and can easily be deleted in error or by physical accident (e.g. storage on top of laser transformers!). Back-up copies of important data and system software should always be kept in a safe place against these eventualities.

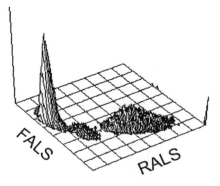

Figure 16 An isometric plot (data as in *Figure 13*).

4.8 Interpretation of acquired data

The results from any instrumentation will always be subject to a degree of variation. Flow cytometric analysis, although capable of measurements with small variance, is also prone to the introduction of systematic errors. Poor sample preparation and staining and improper operation of the instrument leads to inaccurate results. Immunofluorescence measurements become inaccurate in the presence of large numbers of dead cells as these cells take up the antibody non-specifically. If staining is not specific and homogeneous, variation will be introduced that may be larger than the effect to be measured. Furthermore, the emission from fluorochromes may change with pH and temperature. Care must be taken to buffer the pH and to allow the temperature to stabilize before measurements are made. It should also be remembered that scatter measurements are proportional to the size of particles only if the refractive index of the particles is the same.

Measurements are degraded by both optical and electronic noise. Scratched or incorrectly chosen filters will lead to optical noise and poor signal discrimination. Generally, the degree of electronic noise is not within the control of the cytometer user, but it should be monitored by the routine use of a standard, e.g. fluorescent microspheres. PMTs will only produce a linear response within a defined voltage range. Below about 300 volts the response is non-linear and unnecessarily high voltages produce excessive noise. If the PMT becomes saturated with light then non-linearity of response can occur. PMT saturation is prevented by the use of neutral-density filters or control of the power output from the laser. The laser itself is a source of noise. Even the best regulated lasers will retain a small degree of ripple on the light output. Operating the laser at high power outputs reduces the contribution of this constant ripple on the output. Laser light fluctuations and ripple are reduced further if a power control circuit is used. A small proportion of the laser output is monitored and the plasma current regulated by negative feedback to maintain a constant light output.

Any degradation in the accuracy of measurements is detected rapidly if standard particles are used routinely. Fluorescent microspheres are ideal for this purpose, enabling the optical alignment and parameter settings to be reproduced each day. Samples should be analysed only if the standards display acceptable values. However, microspheres are artificial particles and the best settings for these may be inappropriate for biological specimens. If possible, known biological standards or fixed samples should also be run routinely.

A more detailed discussion of quality control will be found in Chapter 8.

5 Cell sorting (see Chapter 4)

The ability of some flow cytometers to separate physically (sort) cells or organelles specifically identified during analysis adds a further dimension to the capability of these instruments. A simple, but very useful, application of flow sorting is the sorting of selected cells on to microscope slides for morphological staining and light microscopy. Important information about the cell types in a complex mixture, particularly when the sample has not previously been analysed, can be gained in this way.

Although it is possible to analyse many cellular parameters directly, it may be necessary to determine cell function by other assay systems. Cell sub-populations can be identified and sorted using flow cytometry before their use in function assays. Flow sorters are particularly adept at identifying and purifying rare events, and have been used for the selection and direct cloning of transfected cells and hybridomas and for the isolation of fetal cells from maternal blood. The purification of specific chromosomes is an important use of flow sorting, and the DNA libraries generated from this material have had a major role in the Human Genome Project.

While other sorting systems have been developed—such as panning or magnetic microbead separation—these are generally single parameter methods, i.e. positively stained cells can be separated from negatively stained cells. Flow sorting has the major advantage that any combination of any analytical parameters can be used to set the criteria for sorting. For example, it is straightforward to define a population of cells which possess the forward and orthogonal scattering characteristics of lymphocytes and are bright for one antigen, dim for a second antigen, and negative for a third and to sort these cells into a tube as a highly purified fraction. It is also possible to generate even more sophisticated sorting gates by the application of other Boolean operators to combinations of sorting gates. For example, a population of cells could be sorted which is dim for one antigen OR is bright for a second antigen but does NOT possess the scattering characteristic of monocytes. By the combination of analytical gates in this way extremely specific sorting criteria can be generated.

References

1. Melamed, M. R., Lindmo, T., and Mendelsohn, M. L. (ed.) (1990). *Flow cytometry and sorting* (2nd edition). Wiley, New York.
2. Shapiro, H. M. (1988). *Practical flow cytometry* (2nd edition). Alan R. Liss, New York.
3. Van Dilla, M. A., Dean, P. N., Laerum, O. D., and Melamed, M. R. (ed.) (1985). *Flow cytometry: instrumentation and data analysis*. Academic Press, London.
4. Watson, J. V. (1991). *Introduction to flow cytometry*. Cambridge University Press, Cambridge.
5. Kelley, K. A. and McDowell, J. A. (1988). *Cytometry*, **9**, 277.

Chapter 2
Fluorescence and fluorochromes

Michael G. Ormerod

34, Wray Park Rd, Reigate, Surrey RH2 ODE, UK

1 Introduction

Fluorescence occurs when a molecule excited by light of one wavelength returns to the unexcited (ground) state by emitting light of a longer wavelength. The exciting and emitted light, being of different wavelengths, can be separated from one another using optical filters. Fluorescence has been used to visualize certain molecules (and hence structures) by light microscopy for many years. It is a sensitive technique because a positive signal is detected against a negative background. It has found widespread application in flow cytometry and, together with light scatter, is the major type of measurement made in most machines. The ability to detect fluorescence simultaneously from two, three, or four compounds fluorescing at different wavelengths enables several parameters to be measured and opens up the field of multi-parametric analysis.

The types of fluorescent probe employed in flow cytometry can be divided into two broad categories: those that are used to label covalently other probes (often antibodies), and those whose distribution or fluorescence reflects their environment and hence particular properties of the cell. Amongst the former, fluorescein is widely used; amongst the latter, fluorescent compounds that bind to DNA are important.

Many companies sell a wide variety of fluorescently labelled antibodies—a list of some of them is given in Appendix 1. For other probes, Molecular Probes Inc., who have specialized in this field, supply the most comprehensive range. Their handbook is also an invaluable source of information.

After outlining the principles underlying fluorescence, this chapter describes the properties of the most commonly used fluorochromes. Details of the use of these compounds will be found in the subsequent chapters.

2 Principles

When a compound absorbs light (and hence energy) electrons are raised from the ground state to an excited state. They return to the ground state by a variety of transitions which, in some compounds, may involve the emission of a quantum of light (radiative transition), see *Figure 1*. The emitted light will always be of lower energy, and hence longer wavelength, than the exciting light. The

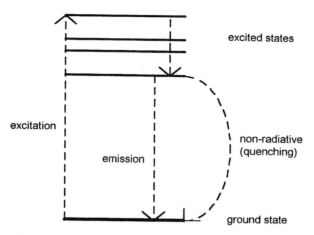

Figure 1 The absorption and emission of light during fluorescence.

energy may also be lost by non-radiative processes and eventually be dissipated as heat.

The amount of light absorbed at a given wavelength is called the extinction coefficient. It is useful if the wavelength of the absorption maximum is close to that of one of the spectral lines of a laser used in flow cytometry (particularly the major line of the argon-ion laser, 488 nm). This is one reason for the popularity of fluorescein, which has an absorption maximum at 495 nm and an extinction coefficient at 488 nm of $8 \times 10^4 \, \text{cm}^{-1} \, \text{M}^{-1}$.

The quantum efficiency is the number of photons emitted for every photon absorbed. This can be affected by the environment; the polarity of the solvent being one such factor. For example, the fluorescence from propidium iodide in aqueous solution increases strongly when the molecule is bound by intercalation to double-stranded nucleic acid. The fluorescence of some molecules, such as fluorescein, is sensitive to pH.

If a fluorochrome interacts closely with another molecule then fluorescence may be quenched, the excitation energy being dissipated by non-radiative transitions. This phenomenon has been used to detect 5-bromodeoxyuridine incorporated into DNA (see Chapter 11). DNA stained with one of the *bis*-benzimidazoles, Hoechst 33342 or Hoechst 33258 (see Section 5), fluoresces blue under UV. Interaction of the bromine atom with the Hoechst dye reduces the fluorescence.

Another important phenomenon is energy transfer. As a result of interaction between two fluorochromes, energy absorbed by one is transferred by a non-radiative transition to the other, which fluoresces (see *Figure 2*). An example of this is observed if fixed cells are incubated with a combination of propidium iodide and the dimeric dye, TO-PRO-3, both of which bind to DNA. When the propidium is excited by an argon-ion laser tuned to 488 nm, the fluorescence wavelength is that of TO-PRO-3 (deep red, 660 nm) as opposed to that of propidium (red, 620 nm) (see Chapter 9, Section 3.2).

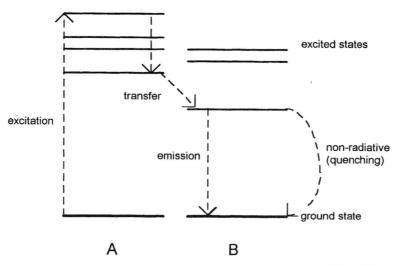

Figure 2 Fluorescence energy transfer. Light is absorbed by molecule A and is transferred to molecule B, which fluoresces.

The lifetime of the excited state is also important since, if it is sufficiently short, a molecule may be excited several times as it passes through the laser beam of a flow cytometer.

Lasers, which are frequently used as the light source in flow cytometers, emit linearly polarized light. Conceptually this can be thought of as the light waves being in a single plane. A fluorochrome will only absorb light if it is correctly oriented relative to the plane of polarization of the laser light. If the fluorochrome is immobilized, the emitted light will also be plane-polarized; if it is tumbling rapidly compared to the lifetime of the excited state, the emitted light will be randomly polarized. This effect has been used to explore the environment of the fluorochrome.

The wavelength of fluorescence needs to be sufficiently removed from that of the exciting light to allow good separation of the two wavelengths by optical filters. (The difference between the absorption and emission maxima is referred to as the Stoke's shift.) When several labels are used, careful choice of fluorochromes is essential to avoid fluorescences that overlap too closely. The number of different labels available has been extended in recent years so that it is possible to measure four fluorescences simultaneously using a single laser. The range can be extended by using two or three lasers.

3 Fluorochromes used to label covalently other probes

The probe to be labelled is commonly an antibody, although it could be another protein such as a lectin, hormone, avidin or streptavidin, or even cDNA. There is a range of fluorescent molecules used to label proteins; the structures of

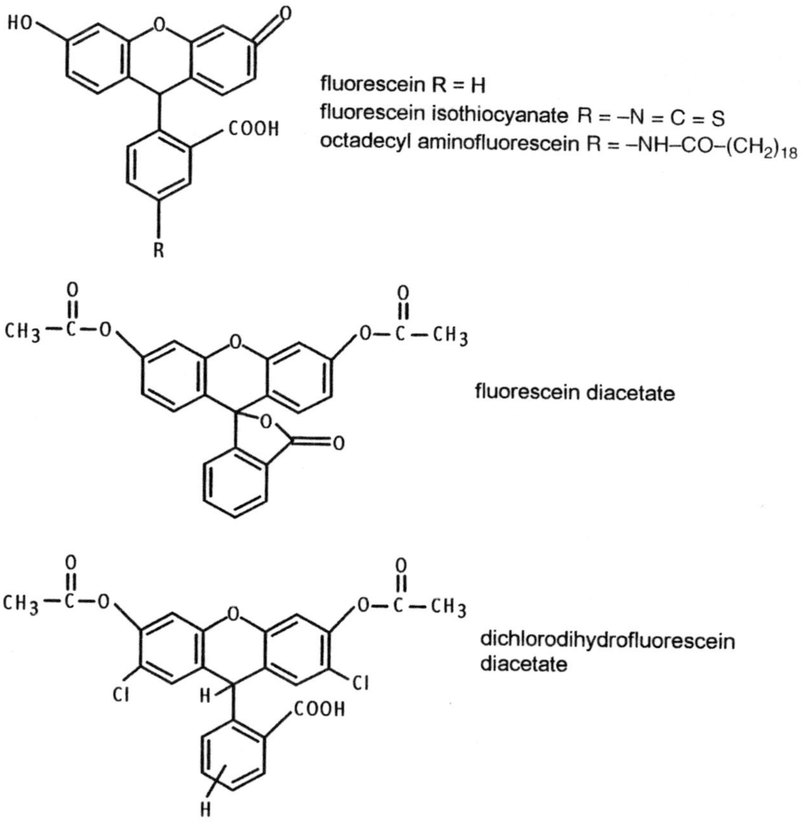

fluorescein R = H
fluorescein isothiocyanate R = –N = C = S
octadecyl aminofluorescein R = –NH–CO–(CH$_2$)$_{18}$

fluorescein diacetate

dichlorodihydrofluorescein diacetate

Figure 3 Fluorescein and some of its derivatives.

fluorescein and some of its derivatives are shown in *Figure 3*. The major properties of some of these fluorochromes are given in *Table 1*.

The fluorochrome needs to be synthesized in a form that can be covalently linked to a protein. An isothiocyanate is commonly used and this can be reacted with the amino groups on the lysine residues of a protein. For this reason, the abbreviation for fluorescein isothiocyanate (FITC) is commonly used, incorrectly, as an abbreviation for fluorescein.

In general, it is best to purchase labelled reagents. Proteins should only be labelled in the laboratory if no commercial alternative is available. Protocols for labelling proteins can be found on Mario Roederer's Web site at www.drmr. com/abcon/.

The most commonly used fluorochrome for light microscopy and flow cytometry is fluorescein (see *Figure 3*). Its advantages are its high extinction coefficient and quantum efficiency, an absorption maximum conveniently close to emission lines from an argon-ion laser and a mercury arc lamp, and its widespread availability linked to a variety of antibodies and other probes.

Table 1 Some fluorochromes used to label proteins

Fluorochrome	Excitation maxima (nm)	Emission maximum (nm)	Laser[a] line (nm)
Fluorescein	495	520	488 A
Biodipy-FL	503	512	488 A
Alexa Fluor Green	491	515	488 A
R-phycoerythrin	564, 495	576	488
Phycoerythrin–Texas Red conjugate	495	620	488 A
Phycoerythrin–cyanine5 conjugate	495	670	488 A
Phycoerythrin–cyanine7 conjugate	495	755	488 A
Peridinin–chlorophyll protein	490	677	488 A
Allophycocyanin	650	660	630 H
Allophycocyanin–cyanine7 conjugate	650	755	630 H

[a]A = Argon-ion, H = helium–neon. Only the larger argon-ion lasers can be tuned to wavelengths other than 488 nm.

It should be noted that the exact wavelengths of absorption and emission may depend on the nature of the conjugate used and its environment.

Molecular Probes have introduced two alternative series of fluorescent reagents. One of these is based on 4,4-difluoro-4-bora-3a, 4a-diaza-s-indacene (BIODIPY). BIODIPY-FL, fluoresces green upon excitation at 488 nm but has a higher quantum efficiency and narrower emission bandwidth than fluorescein. It is also insensitive to pH. The other series is based on sulfonated rhodamine derivatives (trade name, Alexa Fluor). Alexa Green, a substitute for fluorescein, has a higher absorption coefficient, higher quantum efficiency, and better photostability than fluorescein; its fluorescence is insensitive to pH. To date, these reagents have found limited use, probably because the number of antibody conjugates available is still small.

When a second label is required, the label of choice is R-phycoerythrin (PE). This fluorescent protein can be excited at 488 nm, so that only one laser is required, and it emits orange light, which can be easily distinguished from the green fluorescence of fluorescein.

There are several reagents now available for a third label which can be excited at 488 nm and which gives red fluorescence. Peridinin chlorophyll-A protein (PerCP) and the phycoerythrin–cyanine5 conjugate (Cy-Chrome or PC5) both fluoresce deep red (680–690 nm). A conjugate of phycoerythrin and Texas Red (ECD) fluoresces red (620 nm).

In an experiment involving four immunofluorescent labels, fluorescein, PE, ECD, and either PerCP or PE–Cy5 or PE–Cy7 can be used. In a flow cytometer equipped with an additional helium–neon laser (giving a laser line at 630 nm), allophycocyanin and allophycocyanin–Cy7 conjugate could be used allowing at least five immunofluorescences to be recorded.

Details of immunofluorescence analysis can be found in Chapter 5.

4 Fluorochromes used to label directly cell components

4.1 Loading dyes into cells

Some of the compounds described below (see Sections 4.5–7) are used in viable cells. However, many are charged and will not readily diffuse into cells. Often the compound can be obtained as an uncharged acetyl or acetoxymethyl ester, which will freely cross the plasma membrane. Inside the cell, esterases remove the acetyl or acetoxymethyl groups releasing the fluorescent parent compound. The charged compound will remain trapped in the cell. The structure of fluorescein diacetate, which can be used to introduce fluorescein into cells, is given in *Figure 3*.

4.2 Probes for nucleic acids (see *Table 2*)

The probes described in this section, unless stated otherwise, bind stoichiometrically to nucleic acids so that they can be used for quantitative measurement. This is particularly important for the measurement of cell ploidy and the visualization of the cell cycle (see Chapter 6).

With the exception of some of the *bis*-benzimidazoles, such as Hoechst 33342, and the laser dye, LDS 751, the compounds listed in *Table 2* do not freely cross an intact plasma membrane and cells have to be fixed or permeabilized before staining for DNA.

The phenanthridiniums, propidium and ethidium, intercalate between the

Table 2 Some fluorochromes used to label nucleic acids

Fluorochrome	Excitation maxima (nm)	Emission maximum (nm)	Laser line (nm)
Propidium iodide	495, 342	639	488 A
Ethidium bromide	493, 320	637	488 A
7-Aminoactinomycin D	546	647	514 or 488 A
Acridine Orange	503	530 (DNA) 640 (RNA)	488 A
Chromomycin A_3	430	580	457 A
Hoechst 33342	395	450	UV A or HeCd
DAPI	372	456	UV A or HeCd
Pyronin Y	545	565	514 or 488 A
Thiazole Orange	509	533	488 A
YO-PRO-1	491	509	488 A
TO-PRO-3	642	661	630 H
LDS 751	543	712	488 A

[a] A = Argon-ion, HeCd = helium–cadmium

The wavelengths above are those after binding to nucleic acid since the fluorescent properties of most of these dyes change on binding. See also notes to *Table 1*.

bases in double-stranded nucleic acids. They may be excited either by UV or blue light giving red fluorescence. With an argon-ion laser tuned to 488 nm, they can be used in combination with fluorescein, which makes them particularly useful for the simultaneous measurement of antibody binding and DNA content (see Chapter 9). Propidium and ethidium also bind to double-stranded RNA which has to be removed if DNA is to be measured (Chapter 6).

Acridine Orange fluoresces green when intercalated in double-stranded nucleic acids. It has a different mode of binding to single-stranded nucleic acids whereby it stacks on the charged phosphates; in this form it fluoresces red. These properties have been exploited to measure DNA (double-stranded) and RNA (single-stranded) simultaneously and also to investigate the acid-induced denaturation of DNA (1). A related compound, pyronin Y, is used to label RNA after its binding sites to DNA have been blocked by a non-fluorescent dye or the DNA removed by enzymatic treatment.

The *bis*-benzimidazoles bind preferentially to AT-rich regions in the small groove of DNA and fluoresce blue when excited by UV light. However, at high concentrations, there is a second mode of binding that causes the fluorescence spectrum to shift towards the red as the concentration of bound dye increases above a critical value. Hoechst 33342 is also actively transported in and out of cells so that the concentration of dye in a cell and its fluorescent spectrum will depend on the conditions of incubation. These phenomena can be used to distinguish between different cells in a mixture (2). Active transport can also be observed with some other drugs that bind to DNA.

The structurally related antibiotics, mithramycin and chromomycin A_3 bind to the GC-rich regions of DNA. Chromomycin is used in conjunction with Hoechst 33342 to stain chromosomes (see Chapter 12).

Thiazole Orange is used to stain RNA in reticulocytes (see Chapter 7).

The laser dye, LDS 571, is cell permeant. While its binding is not stoichiometric and does not allow cell-cycle analysis, it can be used to delineate nucleated cells. It will bind more strongly to damaged cells and this difference persists after fixation in paraformaldehyde, allowing a dead/live discrimination to be made (3).

A new set of cyanine-based nucleic acid stains has been introduced by Molecular Probes Inc. (4). They have exceptional sensitivity and cover a wide range of emission wavelengths. The spectral properties of two of the compounds are given in *Table 2*.

4.3 Probes that reflect membrane potential (see *Figure 4* and *Table 3*)

The distribution of some dyes reflects the electrical potential difference across the relevant membrane (see Chapters 13 and 14). The most commonly used are the family of cyanine dyes, whose general structure is given in *Figure 4*, and the oxonol dyes (5).

Cationic dyes label mitochondria within the cell strongly, presumably because

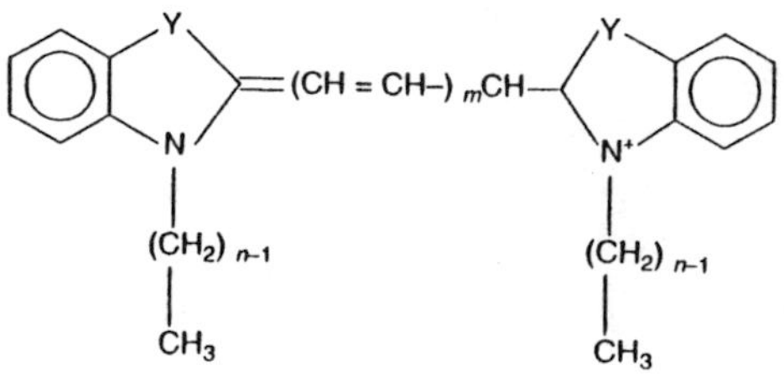

Figure 4 The structure of cyanine dyes. The variable substituent, Y, may be oxygen (O), sulfur (S), or isopropyl (I).

Table 3 Some fluorochromes used as probes for membrane potential

Fluorochrome	Excitation maximum (nm)	Emission maximum (nm)	Laser line (nm)
DiOC$_6$(3)	484	501	488 A
DiOC$_5$(3)	484	501	488 A
di-BA-C$_4$(3)	543	590	488 or 514 A
DiSBaC$_2$(3)	535	560	514 A
Rhodamine 123	485	546	488 A
MitoTracker Red CMXRos	578	599	488 or 514 A
JC-1			
–monomer	510	527	488 A
–dimer	~585	590	488 A
Merocyanine 540	500	572	488 A

See notes to *Table 2.1*

these organelles have a relatively high negative membrane potential. In particular, rhodamine 123, 3,3′-dihexyloxacarbocyanine (DiOC$_6$(3)), a chloromethyl derivative of X-rhosamine (MitoTracker Red CMXRos) and 5,5′,6,6′,-tetrachloro-1,1′,3,3′-tetraethylbenzimidazolylcarbocyanine iodide (JC-1) have been used for this purpose (6) (see Chapter 14, Section 6). JC-1 fluoresces green, but at high concentrations it aggregates and both the absorption and emission spectra shift to longer wavelengths and the dye complexes fluoresce red. In cells, aggregate formation increases linearly with increasing membrane potential. Changes in membrane potential in mitochondria can be followed by observing changes in the ratio of red:green fluorescence.

Merocyanine 540 appears to bind preferentially to loosely packed membranes with highly disordered lipids, and has been used to follow changes in the structure of the plasma membrane during lymphocyte differentiation (7) and during apoptosis (8).

4.4 Probes for lipids (see *Figure 7* and *Table 4*)

These probes are either uncharged hydrocarbons or because of their long alkyl side chains, are soluble in lipids. The way in which different probes are partitioned between the membranes in cells depends on their structure. For

Table 4 Some other useful fluorochromes

Fluorochrome	Excitation maxima (nm)	Emission maximum (nm)	Laser line (nm)
Nile Red	485	525	488 A
PKH26	551	567	488 or 514 A
CM-DiI	553	570	488 or 514 A
DiOC18(3)	484	501	488 A
CMTMR	541	565	488 or 514 A
Indo-1	349	400/475[a]	UV A or HeCd
Fluo-3	503	526	488 A
Fura Red	436/472	637/657 [b]	488 A
SNARF 1	514	580/630 [b]	488 or 514 A
Rhodamine 110	499	521	488 A
MitoTracker Green FM	490	516	488 A

[a]Dependent on Ca^{2+} concentration

[b]Dependent on pH

See notes to *Tables 2.1* and *2.2*.

example, Nile Red is specific for lipid oil droplets (9) while C18-fluorescein (10), C18-cyanines, such as octadecyl indocarbocyanine, CM-DiI, (11) or PKH26 (12) can be used as general purpose labels for the plasma membrane (see Chapter 7, Section 8).

4.5 Probes sensitive to calcium (see *Table 4* and Chapter 13)

Cytoplasmic calcium plays an important role in regulating several cellular reactions, and fluorescent dyes have been developed whose properties are sensitive to the concentration of Ca^{2+}. They are usually used as (non-fluorescent) acetoxymethyl esters (Section 4.1).

Quin-2, the original dye used for this purpose, and its more sensitive analogue, fura-2, respond to an increase in Ca^{2+} concentration by a shift in the absorption spectrum towards longer wavelengths. For flow cytometry, if a UV laser is available, the dye, indo-1, is preferred because it is the emission spectrum which changes. The dye is excited with UV and the fluorescence either side of the emission maximum is ratioed (see Chapter 13). For use in instruments equipped only with a laser producing light at 488 nm, changes in the emission intensity are measured using either fluo-3 or Fura Red; sometimes both probes are used in combination (Chapter 13).

4.6 Probes used for long-term cell labelling (see *Table 4*)

There are two types of dye used to label cells for long-term experiments. In one, the fluorescent probe is attached to a lipophilic side chain that inserts into the plasma membrane. The other dyes are trapped within the cell. Examples of the first type are PKH26, an analogue of Acridine Orange with an N-linked, 26-carbon alkyl chain (12), octadecafluorescein (10), and CM-DiI, a chloromethyl derivative of a carbocyanine with alkyl side chains (11). The chloromethyl group reacts with protein thiols stabilizing the dye in the plasma membrane.

5-Chloromethylfluorescein diacetate (CMFDA) is a tracker dye that is trapped within the cell. CMFDA diffuses freely into cells where it is converted by esterases into 5-chloromethylfluorescein. It is then stabilized within the cell by reaction with protein thiols. The dye remains in the cytoplasm several days, if not weeks, after labelling (13); as does carboxyfluorescein diacetate, succinimidyl ester (CFDA SE).

Other dyes which have been used for cell-fusion experiments, where long-term stability is not as important, include 3,3'-dioctadecyloxacarbocyanine, $DiOC_{18}(3)$, octadecyl fluorescein (10) and CMFDA, and 5 (and 6)-([(4-chloromethyl)benzoyl]amino) tetramethylrhodamine (CMTMR) (14).

4.7 Other probes (see *Table 4*)

Molecular Probes Inc. have introduced a series of dyes—seminaphthorhodafluors (SNARF) and seminaphthofluoresceins—whose emission and excitation spectra are sensitive to pH. For example, the compound referred to by Molecular Probes as SNARF-1 can be excited at 488 nm and has an emission spectrum whose maximum shifts from 580 nm to 640 nm when the pH changes from 6 to 9. The use of these dyes is discussed in Chapter 13.

Another interesting type of probe is non-fluorescent but can be oxidized to give a fluorescent product. This property can be used to follow the production of oxidative species intracellularly when, for example, neutrophils undergo a respiratory burst upon stimulation. Dyes in this category include dichlorofluorescin (oxidized to dichlorofluoroscein) (see *Figure 3*) (15), dihydrorhodamine 123 (oxidized to rhodamine 123 which stains mitochondria) (16), and dihydroethidium (oxidized to ethidium which stains DNA) (17). Dihydrorhodamine and dihydroethidium diffuse into cells. The other compounds described above, based on fluorescein, are charged and are loaded into cells as acetyl esters (see Section 4.1).

The production of a fluorescent compound by enzymatic action within the cell can be used to monitor the concentration of a variety of enzymes. The production of fluorescein from fluorescein diacetate is a particular example. Substrates specific for individual proteases have been produced by linking an appropriate tri- or tetra-peptide to rhodamine 110 (18) (see *Figure 5*). These substrates have been used to observe the activity of caspases, enzymes central to the apoptotic cascade (see Chapter 14).

One particularly interesting enzyme substrate is fluorescein di-β-D-galactopyranoside from which fluorescein is produced by the enzyme, β-galactosidase—the product of the reporter gene, *Escherichia coli lacZ*. Reporter genes are fused with other genes, or with genomic regulatory elements, and the resulting DNA constructs introduced into cells. The reporter-gene product can be used to

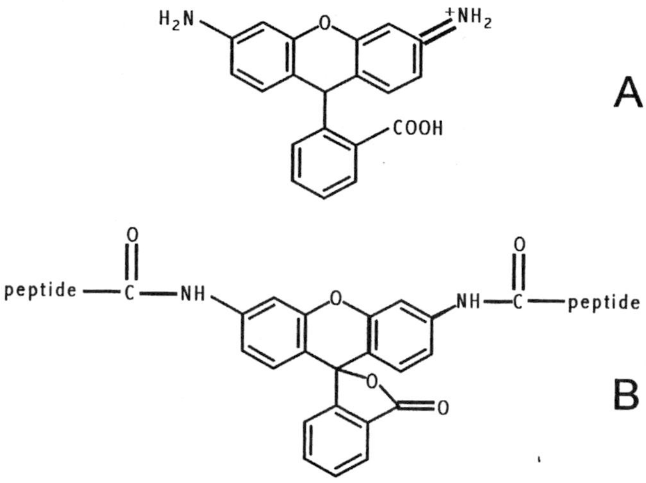

Figure 5 Rhodamine 110 (A) and its derivative (B) used to measure protease activity in cells.

measure the incorporation of the construct and to select these cells by sorting. As the product is an enzyme, a flow cytometric method for its detection allows measurements on individual cells and also permits the selection of cells by cell sorting (19). A detection kit is available commercially and includes a comprehensive protocol for cell labelling (from Molecular Probes Inc.).

Another reporter molecule, now in widespread use, is green fluorescent protein. The gene for this protein is used in the same way as that for *lacZ*. The cells that have incorporated the gene construct can be detected by their green fluorescence (20, 21).

There are probes that are taken up by mitochondria but which are mitochondrial membrane-potential insensitive. Such probes can be used to estimate mitochondrial mass. One of these (described in Chapter 14, Section 4) has been given the trade name MitoTracker Green FM.

References

1. Darzynkiewicz, Z. and Kapuscinski, J. (1990). In *Flow cytometry and cell sorting* (ed. M. R. Melamed, T. Lindmo, and M. L. Mendelsohn), p. 291. Wiley-Liss, New York.
2. Watson, J. V., Nakeff, A., Chambers S. H., and Smith, P. J. (1985). *Cytometry*, **6**, 310.
3. Terstappen, L. W. M. M., Shah, V. O., Conrad, M. P., Rectenwald, D., and Loken, M. R. (1988). *Cytometry*, **9**, 477–84.
4. Haughland, R. P. (1996). *Handbook of fluorescent probes and research chemicals* (6th edn). Molecular Probes, Inc., Oregon.
5. Waggoner, A. S. (1990). In *Flow cytometry and cell sorting* (ed. M. R. Melamed, T. Lindmo, and M. L. Mendelsohn), p. 209. Wiley-Liss, New York.
6. Cossarizza, A., Kalashnikova, G., Grassilli, E., Chiappelli, F., Salvioli, S., Capri, M., Barbieri, D., Troiano, L., Monti, D., and Franceschi, C. (1994). *Exp. Cell Res.*, **214**, 323.
7. Del Buono, B. J., Williamson, P. L., and Schlegel, R. A. (1986). *J. Cell. Physiol.*, **126**, 379.

8. Fadok, V. A., Voelker, D. R., Cammpbell, P. A., Cohen, J. J., Bratton, D. L., and Henson, P. M. (1992). *J. Immunol.*, **148**, 2207.

9. Greenspan, P., Mayer, E. P., and Fowler, S. D. (1985). *J. Cell Biol.*, **100**, 965.

10. Gant, V. A., Shakoor, Z., and Hamblin, A. S. (1992). *J. Immunol. Meth.*, **156**, 179.

11. Andrade, W., Seabrook, T. J., Johnston, M. G., and Hay, J. B. (1996). *J. Immunol. Meth.*, **194**, 181.

12. Teare, G. F., Horan, P. K., Slezak, S. E., Smith, C., and Hay, J. B. (1991). *Cell. Immunol.*, **134,** 157.

13. Zhang, Y.-Z., Kang, H. C., Kuhn, M., Roth, B., and Haugland, R. P. (1992). *Mol. Biol. Cell*, **3**, 90a.

14. Jaroszeski, M. J., Gilbert, R., and Heller, R. (1994). *Anal. Biochem.* **216**, 271.

15. Bass, D. A., Parce, J. W., Dechatelet, L. R., Szejda, P., Seeds, M. C., and Thomas, M. (1983). *J. Immunol.*, **130**, 1910.

16. Rothe, G., Oser, A., and Valet, G. (1988). *Naturwiss.*, **75**, 354.

17. Rothe, G. and Valet, G. (1990). *J. Leucocyte Biol.*, **47**, 440.

18. Assfalg-Machleid, I., Rothe, G., Klingel, S., Banati, R., Mangel, W. F., Valet, G., and Machleid, W. (1992). *Biol. Chem. Hoppe-Seyler*, **373**, 433.

19. Fiering, S. N., Roederer, M., Nolan, G. P., Micklem, D. R., Parks, D. R., and Herzenberg, L. A. (1991). *Cytometry*, **12**, 291.

20. Galbraith, D. W., Anderson, M. T., and Herzenberg, L. (1999). In *Methods in Cell Biology*, Vol. 58 (ed. K. F. Sullivan and S. A. Kay), p. 315. Academic Press, San Diego, CA.

21. Ropp, J. D., Donahue, C. J., Wolfgang-Kimbell, D., Hooley, J. J., Chin, J. Y. W., Hoffman, R. A., Cuthbertson, R. A., and Bauer, K. D. (1995). *Cytometry*, **21**, 309.

Chapter 3
Preparing suspensions of single cells

Collated by Michael G. Ormerod
34, Wray Park Rd, Reigate, Surrey RH2 0DE, UK

1 Introduction

A suspension of single cells is the essential element of flow cytometry. The sample should contain as little debris and as few dead cells and clumps as possible. A set of protocols for the preparation of cell suspensions has been collected in this chapter. It is by no means comprehensive but should form a good starting point for any new investigation. Extensive sets of methods have also been described by Cheetham *et al.* (1) and Pallavicini (2).

The first six protocols have been taken from N. P. Carter's Chapter 3 in the first edition of this book.

2 Cells from peripheral blood

2.1 Introductory remarks

Peripheral blood contains a suspension of single cells ideal for flow cytometry. Freshly taken blood must be prevented from clotting by defibrination or by the use of tubes containing either disodium EDTA (2 mg/ml), sodium citrate (0.38%), or heparin (15 IU/ml). Preparation then entails partial purification of the cell type to be studied.

Protocol 1 describes the preparation of some commonly used reagents.

Protocol 1

Standard reagents

A. 10% bovine serum albumin (BSA)

1. Dissolve 10 g of BSA (Cohn Fraction V, Sigma) in 100 ml distilled water.

2. Centrifuge at 20 000 g for 30 min at 4 °C.

3. Aliquot the supernatant into appropriate amounts and store at −20 °C.

Protocol 1 continued

B. 0.1% BSA–DAB solution

1. Prepare 500 ml of Dulbecco's phosphate-buffered saline (DAB), pH 7.3 (Sigma or Gibco BRL).
2. Add 5 ml 10% BSA.
3. Pass through a 0.45 μm filter.

C. Ammonium chloride lysis buffer

1. Dissolve 8.29 g NH_4Cl, 1 g $KHCO_3$, and 37 mg Na_2EDTA in 1 litre of distilled water.
2. Adjust the pH to 7.2.
3. Make up fresh lysis buffer each time and pass through a 0.45 μm membrane filter before use.

2.2 Platelets

Platelets are much smaller in size and have a lower sedimentation rate than other cells in peripheral blood. They can be purified by centrifugation.

Protocol 2

Preparation of platelets

Reagents

- Heparinized peripheral blood
- BSA–DAB (see *Protocol 1*)

Method

1. Centrifuge 10 ml of heparinized peripheral blood at 400 g for 10 min at room temperature.
2. Remove platelet-rich supernatant and centrifuge at 1000 g for 10 min to obtain a platelet pellet.
3. Wash the pellet twice with BSA–DAB and resuspend to the appropriate concentration.

2.3 Preparation of leucocytes

Leucocytes comprise only approximately 0.1% of the cells in peripheral blood and they need to be enriched. Two methods are used. The first uses an ammonium chloride buffer to lyse the red cells and the ghosts are then removed by washing (see *Protocol 3*). This method is often combined with a washing step during an antibody labelling procedure (see *Protocol 4*, Chapter 5). There are also several commercial reagents available for use with cell-surface markers, some of which also fix the leucocytes as well as lysing the red cells.

The second method uses centrifugation over a step gradient of a mixture of Ficoll and sodium diatrizoate (Histopaque) to select a particular type of leucocyte according to its density (*Protocol 4*). Ficoll–sodium diatrizoate of specific gravities 1.119 and 1.077 can be used in a double gradient to enrich granulocytes and mononuclear leucocytes separately from the same blood sample. The granulocytes sediment to the 1.119/1.077 interface while the mononuclear cells are found at the 1.077/plasma interface.

Protocol 3

Preparation of leucocytes by lysis of red blood cells

Equipment and reagents
- Anticoagulated blood
- 50 ml centrifuge tubes
- NH$_4$CL lysis buffer (see *Protocol 1*)
- BSA–DAB (see *Protocol 1*)

Method
1. Collect 10 ml of anticoagulated blood. Aliquot 2 ml into each of five 50 ml centrifuge tubes.
2. To one tube at a time, rapidly add 45 ml of NH$_4$CL lysis buffer and mix immediately.
3. Incubate at room temperature for 10 min.
4. Centrifuge at 400 g for 10 min at 4°C. Decant the supernatant. Add 5 ml of BSA–DAB to each pellet and combine into one tube.
5. Centrifuge at 400 g for 10 min at 4°C. Decant the supernatant and wash with a further 25 ml of BSA–DAB.
6. Adjust the cell concentration to 5×10^6 cell/ml and store on ice.

Protocol 4

Preparation of human mononuclear cells using density-gradient separation

Equipment and reagents
- Anticoagulated blood
- PBS, pH 7.3
- Histopaque 1077 (Sigma)
- BSA–DAB (see *Protocol 1*)

Method
1. Collect 10 ml of anticoagulated blood. Dilute with an equal volume of PBS.
2. Underlay with 5 ml Histopaque 1077 (Sigma).
3. Centrifuge at 700 g for 30 min at room temperature.

Protocol 4 continued

4. Carefully aspirate the opaque layer of mononuclear cells from the plasma/ Histopaque interface.

5. Make up to 20 ml with BSA–DAB and centrifuge at 400 g for 10 min at 4°C.

6. Wash twice with 10 ml BSA–DAB at 400 g for 5 min at 4°C.

7. Resuspend the final pellet to a cell concentration of 5×10^6 cells/ml.

3 Leucocytes from other tissue

3.1 Lymphoid tissues

Lymph nodes, tonsil, spleen, and thymus are relatively easy to break up into a suspension of single cells. The tissue is chopped coarsely, teased apart, and filtered.

Protocol 5

Preparation of a single-cell suspension from lymphoid tissues

Equipment and reagents
- BSA–DAB (see *Protocol 1*)
- Two scalpels or pairs of forceps
- Fine mesh, stainless-steel tea strainer
- Histopaque (see *Protocol 4*)

Method

1. Place the tissue in a Petri dish with 15 ml of BSA-DAB. Cut into small portions 3–4 mm in size and carefully tease apart each piece with two scalpels or pairs of forceps.

2. Pass through a fine mesh, stainless-steel tea strainer and then wash through a single sheet of lens cleaning tissue.

3. Underlay with Histopaque and proceed as in *Protocol 4*, step 3.

3.2 Other tissues

Generally, other tissue have to be disaggregated by enzymatic digestion. The best conditions have to be worked out for each tissue used. The method described in *Protocol 6* releases infiltrating cells from kidney.

Protocol 6

Preparation of infiltrating cells from rat kidney

(M. Dallman, Nuffield Department of Surgery, Oxford)

Equipment and reagents
- Scalpel blade
- 1 mg/ml collagenase solution (Sigma) in serum-free, tissue culture medium
- CO_2 incubator
- Fine mesh, stainless-steel tea strainer
- Histopaque (see *Protocol 4*)

Method
1. Place the organ in a Petri dish and chop into small pieces with a scalpel blade.
2. Add 10 ml of collagenase solution and incubate at 37°C for 30 min in a CO_2 incubator.
3. Press through a fine mesh, stainless-steel tea strainer and then wash through a single sheet of lens cleaning tissue.
4. Underlay with Histopaque and proceed as in *Protocol 4*, step 3.

4 Other types of cell

4.1 Enzymatic methods

The heterogeneity of other tissue has generated a similar diversity of preparation techniques, many of them involving enzymatic digestion. A technique that works well with one tissue will not necessarily be successful with another. Some other methods used with different solid tissues will be found in references 1–3.

A generalized procedure for obtaining single cells from a tissue by digestion with collagenase is given in *Protocol 7*. A more sophisticated versions of this procedure has been used to isolate epithelial cells from human and rat mammary glands for flow cytometry (4, 5).

Protocol 7

Preparation of cells from a tissue by digestion with collagenase

(G. D. Wilson, CRC Gray Laboratories, Mount Vernon)

Equipment and reagents
- Scalpels
- Hanks' buffered salt solution (HBSS) (Sigma) or PBS, pH 7.3
- HBSS without Ca^{2+} or Mg^{2+}
- Collagenase Type II (Sigma)
- DNase 1 (Sigma)
- 5 ml automatic pipette with plastic tip
- 35 µm nylon mesh (Lockertex Ltd, or Small Parts)
- 0.15% trypsin (Lorne Diagnostics)
- 1% BSA or 5% FBS

Protocol 7 continued

Method

1. Place the tissue in a Petri dish and mince with scalpels into 1 mm³ fragments.

2. Wash in HBSS or PBS and discard the supernatant.

3. Add 10 ml HBSS without Ca^{2+} or Mg^{2+} but containing 0.2% collagenase Type II and 0.02% DNase 1.

4. Incubate at 37°C with constant shaking or rotation for between 15 and 60 min (depending on the nature of the tissue).

5. Take the fragments up and down through a 5 ml automatic pipette with plastic tip to break up the pieces into single cells. Continue until only connective tissue remains.

6. Filter through the 35 μm nylon mesh.

7. Centrifuge at 300 g for 5 min.

8. Resuspend the pellet in PBS or in tissue culture medium.

   With some tissues, step 5 will yield small 'organoids' but it is difficult to obtain single cells. If this is so, proceed as follows (see refs 4 and 5).

9. Resuspend the pellet from step 7 in 0.15% trypsin (Lorne Diagnostics) and 0.02% DNase 1 in HBSS without Ca^{2+} or Mg^{2+}. Incubate at 37°C for a time to be determined by experiment (between 2 and 20 min).

10. Halt the action of trypsin by the addition of 1% BSA or 5% FBS. Proceed as in step 7 above.

The length of incubation time in collagenase (*Protocol* 7, step 4) can be determined by periodically pipetting the tumour pieces to see if they break up. The plastic tip used in *Protocol* 7, step 5 can be cut to a larger diameter to accommodate the pieces of tissue. A narrower tip can be used as disassociation proceeds. Try not to introduce air bubbles during the pipetting as surface tension effects may damage cell membranes.

With some organs, perfusion of the organ *in situ* with an enzyme solution is used. *Protocol* 8 describes the use of this type of method with rat lungs. Hepatocytes have been prepared by a similar technique (6).

Protocol 8

Dispersal of rat lung to a single-cell suspension

(Janet Martin and Ian N. H. White, MRC Toxicology Unit, Leicester)

Equipment and reagents

- Perfusion apparatus (see *Figure 1*)
- HBSS without Ca^{2+} or Mg^{2+} (see *Protocol 7*)
- 7.5% $NaHCO_3$
- 1 mM EDTA
- 95% O_2/5% CO_2 source
- Subtilisin[a]

Protocol 8 continued

- 0.2 μm filter
- 250 mg/kg pentobarbitone
- Artery forceps
- 2 pairs of fine curved forceps
- 1000 U/ml heparin solution
- Bulbous-ended cannula[b]
- 3 mm clip

- Plastic cannula[c]
- PBS, pH 7.3
- Plastic Petri dishes
- Thermistor probe
- FBS
- DNase
- 125 μm mesh, nylon gauze

Method

1. Assemble the perfusion apparatus as shown in *Figure 1*, but without the lungs and tracheal cannula. Use HBSS without Ca^{2+} or Mg^{2+}, and adjust the pH to 7.4 with 7.5% $NaHCO_3$. Add 150 ml HBSS + 1 mM EDTA, pH 7.4, to the beaker which will hold the lungs. Balance the pumps to maintain a constant fluid level in the oxygenator, and make sure the tubing contains no gas bubbles that might enter the lung. The outlet tube from the second pump terminates in a plastic Luer nozzle. Oxygenate with 95% O_2/5% CO_2 at 500 ml/min.

2. Remove 10 ml of the HBSS/EDTA solution from the beaker and use it to dissolve 15 mg subtilisin.[a] Filter this solution through a 0.2 μm filter.

3. Anaesthetize a rat intraperitoneally with pentobarbitone at 250 mg/kg. (Rapid anaesthesia within 5 min minimizes peripheral blood clotting during perfusion.) Wet the ventral surface of the animal with 70% ethanol and lay the rat on an operating table with all four limbs secured.

4. Open the abdominal cavity and loosen and deflect the skin. Cut across the diaphragm close to the bottom ribs. Cut up through both sides of the rib cage and clamp the xiphisternum with artery forceps, leaving the sternum deflected over the head by the weight of the forceps. Carefully cut the pericardium to free the heart, which should still be faintly beating at this stage. Apply 0.5 ml of the 1000 U/ml heparin solution to the surface of the heart and lungs.

5. Attach a bulbous-ended cannula[b] to a syringe containing 2 ml of the heparin solution and insert the cannula through a small incision made in the right ventricle and into the pulmonary artery. Hold the needle in place with a 3 mm clip, clamped at the base of the artery, with only the bulb of the cannula projecting into the artery.

6. Cut the left auricle to allow the perfusate to escape and rapidly, but steadily, inject the heparin solution through the cannula.

7. Remove the syringe and attach the outlet of the perfusion apparatus. Perfuse at 50 ml/min, running the effluent to waste. Maintain vascular perfusion for approximately 5 min.

8. During this time, cut away the liver and cut up both sides of the trachea. Remove the membranes and any tissue overlying the trachea, and tie in a plastic cannula[c] just below the larynx. Ensure the open end of the plastic cannula is in the trachea and do not allow it to enter the bronchi.

Protocol 8 continued

9. Lavage the lungs through this tracheal cannula using three successive 4 ml volumes of PBS.[c] The lungs should ideally be completely blanched after 5 min vascular perfusion and lavage but any small areas of blood clots at the periphery may be cut away after enzyme perfusion.

10. Inflate the lungs with 10 ml PBS with a syringe through the cannula. Holding both cannulae in one hand, with their attached syringe and perfusion tube, cut the heart and lungs free as a single preparation. Remove the clip and the vascular cannula and lay the organs with the tracheal cannula and syringe still attached in a plastic Petri dish. Carefully cut away the heart and as many blood vessels as possible without damaging the trachea. Remove the syringe, leaving the tracheal cannula in place, and allow any PBS to drain out.

11. Once the vascular cannula has been removed, switch the perfusion apparatus over to oxygenate and recirculate 150 ml of HBSS, 1 mM EDTA (the optimum concentration may be less for younger animals) in a 250 ml beaker maintained at 37°C. This can conveniently be allowed to recirculate for a few minutes whilst the organ preparation is being trimmed.

12. Use a syringe to fill the lungs with oxygenated HBSS/EDTA (taken from the 150 ml in the beaker) through the tracheal cannula, with the lungs immersed in the beaker of the same solution. Remove the syringe and attach the perfusion outlet tube to the cannula to start recirculating perfusion. Add the filtered subtilisin solution to the beaker. Continue the recirculating enzyme perfusion for 15 min, checking the temperature is maintained at 37°C by means of a thermistor probe immersed in the beaker of perfusate.

13. Detach the perfusion outlet from the tracheal cannula and allow the perfusate to drain out from the lungs. Flush the lungs, via a syringe attached to the tracheal cannula, with 12 ml HBSS, 10% FBS, 0.05 mg/ml DNase.

14. Remove the lungs to a plastic Petri dish and cut away the cannula and major airways. Shred the lungs in the dish, in the HBSS/FBS/DNase that drains out of the lungs, by shredding the lobes against each other using a pair of closed, fine curved forceps in each hand. Tease the remaining airways free of the lung tissue using the forceps. Continue shredding until all tissue is a maximum size of 1 mm cubes.

15. Further disrupt the tissue by passage (three times) through a 10 ml plastic pipette, followed by filtration through the 125 μm mesh, nylon gauze into a plastic beaker. Wash the remaining tissue/cells from the Petri dish with further HBSS/FBS/DNase and pipette this on to the debris on the gauze. Use further washings to bring the total volume in the beaker to 30 ml.

16. Centrifuge this suspension at 300 g for 4 min at room temperature and aspirate the supernatant to waste. Resuspend the pellet in a further 10 ml of HBSS/FBS/DNase through a pipette and centrifuge as before. Resuspend the resulting pellet in 5 ml HBSS/FBS/DNase and run through a small piece of 125 μm nylon gauze to give the single-cell suspension.

Protocol 8 continued

[a] The choice of dispersal enzyme for lung tissue may depend on the cell type to be isolated. Subtilisin has been the enzyme of choice to give the highest yield of Clara cells. However, trypsin (100 mg in 10 ml) can give good preparations of Type II cells.

[b] Make the cannula made by adding solder around the end of a 4 cm long, 20 gauge needle to give a bulb of 2 mm diameter, extending 3 mm on to the needle.

[c] The cannula should comprise a 5 cm length of flexible plastic tubing with an angled, cut open end attached to a Luer fitting. This can be made by cutting off the winged needle from a Butterfly-19 winged needle infusion set (o.d. 1.1 mm, i.d. 0.8 mm).

[d] Inflation and deflation of the lungs by lavage facilitates clearance of the vascular system. The PBS lavages may be combined to recover alveolar macrophages by centrifuging at 300 g for 4 min at room temperature.

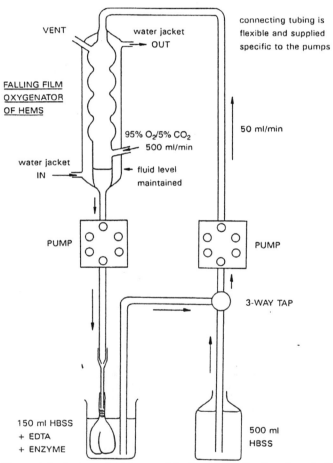

Figure 1 Perfusion apparatus used for the enzymatic digestion of rat lungs. The bottle of HBSS and the beaker to hold the lungs can be secured by clamps in a 37 °C water bath. This bath can also recirculate water through the jacket of the Falling Film Oxygenator of Hems. The system provides for efficient oxygenation of the perfusion solutions, but it could be replaced by a simple gas bubbler.

4.2 A mechanical method

Protocol 9 has been used to prepare cells from solid tumours. The comments after *Protocol 7* are also applicable to this method.

Protocol 9

Mechanical disaggregation of solid tumours

(G. D. Wilson, CRC Gray Laboratories, Mount Vernon)

Equipment and reagents

- Hanks' Buffered Salt Solution (HBSS) or PBS (see *Protocol 7*)
- 5 ml automatic pipette with disposable plastic tip
- Scalpels
- 35 μm nylon mesh (Lockertex Ltd or Small Parts)

Method

1. Place the tissue in a Petri dish and mince with scalpels into 1 mm³ fragments.
2. Wash in HBSS or PBS and discard the supernatant.
3. Add 5 ml HBSS or PBS. Take the fragments up and down through a 5 ml automatic pipette with a disposable plastic tip to break up the pieces into single cells. Continue until only connective tissue remains.
4. Filter through the 35 μm nylon mesh.
5. Centrifuge at 300 g for 5 min.
6. Resuspend the pellet in PBS or in tissue culture medium.

5 Cultured cells

In many cultures of primary cells and of cell lines, the cells grow attached to the plastic surface of the culture flask. It is normal to prepare a single-cell suspension by digestion with trypsin. Care should be taken not to overdigest as this can damage the cells and, in particular, change their surface antigens.

Protocol 10

Preparation of cultured cells

Equipment and reagents

- 0.25% EDTA in PBS, pH 7.2
- 0.25% crystalline trypsin (chymotrypsin-free, Lorne Diagnostics) in PBS, 0.25% EDTA
- Culture medium containing 10% FBS, 1% soybean trypsin inhibitor (Sigma)
- Air-buffered culture medium plus FBS

Protocol 10 continued

Method

1. Pour the culture medium from the culture flask and wash with PBS, 0.25% EDTA.

2. Add enough 0.25% crystalline trypsin in PBS, 0.25% EDTA to cover the bottom of the flask. Incubate at 37 °C for 2–8 min.

3. Monitor progress of the incubation by knocking the flask sharply to loosen the cells and observing them under a inverted microscope. As soon as the majority of cells are detached and single, stop the action of the trypsin by adding culture medium, 10% FBS, 1% soybean trypsin inhibitor.

4. Wash the cells by centrifugation in air-buffered culture medium plus FBS and finally resuspend at a concentration of about 10^6 cells/ml.

Acknowledgements

The author thanks his colleagues for supplying the protocols in this chapter.

References

1. Cheetham, K. M., Shulka, N., and Fuller, B. J. (1998). In *Cell separation—a practical approach* (ed. D. Fisher, G. E. Francis, and D. Rickwood), p. 1. IRL Press at Oxford University Press, Oxford.

2. Pallavicini, M. G. (1987). In *Techniques in cell cycle analysis* (ed. J. W. Gray and Z. Darzynkiewicz), p. 139. Humana Press, Clifton, NJ.

3. Visscher, D. W. and Crissman, J. D. (1994). In *Methods in Cell Biology*, Vol. 41 (ed. Z. Darnzynkiewicz, J. P. Robinson, and H. A. Crissman), p. 1. Academic Press, San Diego, CA.

4. O'Hare, M. J., Ormerod, M. G., Gusterson, B. A., and Monaghan, P. (1991). *Differentiation*, **46**, 209.

5. Dundas, S. R., Ormerod, M. G., Gusterson, B. A., and O'Hare, M. J. (1991). *J. Cell Sci.*, **100**, 459.

6. Davies, R., Cain, K., Edwards, R. E., Snowden, R. T., Legg, R. F., and Neal, G. E. (1990). *Anal. Biochem.*, **190**, 266.

Chapter 4
Flow sorting

Nigel P. Carter* and Michael G. Ormerod[†]
*Sanger Institute, Hinxton Hall, Hinxton, Cambridge CB10 1RQ,UK
[†] 34, Wray Park Rd, Reigate, Surrey RH2 0DE, UK

1 Introduction

Flow sorting enables subpopulations of cells or particles to be separated from the sample suspension with a high degree of purity. The purified fractions (usually >95% pure) are thus made available for morphological examination or use in functional or other assays, so greatly extending the research and diagnostic potential of the flow cytometer. Any combination of parameters can be applied as criteria to decide whether a cell is or is not sorted, thereby harnessing the full analytical potential of the instrument in the separation process. One of the main original applications of flow sorting was the purification of morphologically similar, but functionally distinct, lymphocyte subpopulations using fluorescent staining with polyclonal antibodies. However, the ability to utilize any analytical parameter or combinations of parameters has led to flow sorting being used in diverse applications, ranging from cloning rare cell transfectants directly into multi-well plates to the bulk sorting of human chromosomes for the production of DNA libraries.

It is not within the scope of this chapter to present practical details of flow sorting with specific commercial instruments as each has slightly different modes of operation. While the information presented in this chapter will cover the general principles and practicalities of flow sorting, it will be necessary to refer to instrument manuals for specific operational procedures.

2 Principles of flow sorting

The most widespread method for flow sorting was adapted from technology designed for ink-jet graphic printing and utilizes the electrostatic deflection of droplets. A generalized sorting system is shown in *Figure 1*. In droplet sorting, particles exit the flow chamber in a jet which breaks up into regularly spaced droplets. Specific droplets containing the cells of choice are charged and passed through a high-voltage electrostatic field thus being deflected for collection.

Both quartz-cuvette and 'stream in air' flow chambers (see Chapter 1, Section 3.3.2) can be utilized for droplet sorting. The pressurized sheath fluid exits the flow chamber through an accurately drilled hole (usually with an orifice

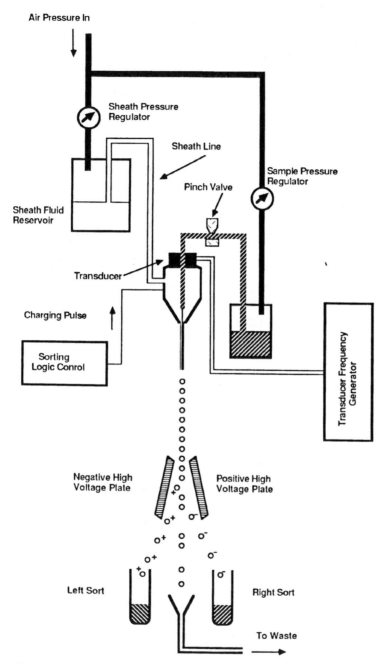

Figure 1 A generalized sorting system.

diameter of between 50 and 100 μm) to form the jet. Any stream of liquid in air has a tendency to break up into droplets but does so in an unpredictable fashion. However, under the influence of a constant vibration of an appropriate frequency, a standing wave is formed in the jet and stable drops, one wave-

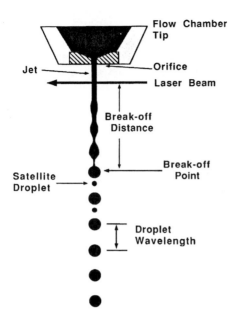

Figure 2 Droplet formation.

length apart, are generated (see *Figure 2*) at a specific distance from the orifice (the break-off point).

The applied vibration is produced by mounting the flow chamber in a piezo-electric transducer to which a regular voltage waveform is applied. The waveform amplitude determines the distance from the nozzle to the break-off point. During droplet formation, the fluid in the neck region close to the break-off point forms into a separate satellite droplet that can be seen under stroboscopic illumination. Within two or so further wavelengths these satellites merge into the main droplets. Application of a voltage pulse to the sheath stream for the period while a droplet is forming results in a net charge on the droplet breaking away, while droplets before and after remain essentially uncharged. The stream of droplets then pass through an electrostatic field formed between high-voltage plates. Charged droplets are attracted to the opposite pole and in this way become separated from uncharged or oppositely charged droplets. Because it is possible to apply either a negative or a positive charge to a droplet, cells can be sorted to the left or the right of the undeflected stream, thereby allowing two subpopulations to be sorted. In one commercial instrument (Cytomation's MoFlo MLS) two different positive or negative voltages can be applied to a droplet, so that four subpopulations can be selected (two sorting to the right, two to the left).

While droplets form at the break-off point, the particles are analysed either in the quartz cuvette above the orifice or immediately below the nozzle in the emerging jet. This means that there is a time delay between the sorting decision and the cell of choice being encapsulated in a charged droplet. Consequently, electronic circuitry is employed to delay the charging pulse for the precise length of time necessary for the cell to travel from the detection point to the jet break-off point. This is known as the drop delay and must be set exactly.

Cells are selected for sorting by the logic control circuitry. The earlier instruments used hard-wired logic circuitry where value ranges for each parameter were set using potentiometers. Cells falling within these ranges were selected for sorting. In practice, this allowed only rectangular regions of interest to be set on cytograms. Today, sorting is controlled by computer using bit-mapped display techniques to set irregular shaped regions. Boolean logic can also be applied on computer systems to allow complex sorting criteria to be set. In modern high-speed sorters, the data transfer between the sorter and computer may be too slow to handle the number of events (>25 000/sec). In one commercial instrument, the sort decisions are loaded from the computer into look-up tables, which map the univariate or bivariate histograms.

3 Parameters which affect sorting

3.1 General considerations

Many parameters influence droplet formation and the efficiency of sorting. These include:

- transducer frequency;
- transducer amplitude;
- orifice diameter;
- sheath pressure and viscosity;
- drop delay;
- charge phase angle;
- anti-coincidence settings.

For accurate sorting, the coincidence of the application of the charging pulse and the arrival of the cell at the break-off point must be exact. This can only be attained if stable, regular droplet formation and a fixed break-off point is achieved. Droplet formation is dependent on the transducer frequency and amplitude, as well as the sheath diameter (orifice size), sheath viscosity, and sheath pressure (jet speed). Regular droplet formation occurs when the wavelength of the applied vibration is greater than π times the jet diameter but, in practice, stable droplet patterns are formed between four times and eight times the jet diameter. Under these droplet formation conditions, the amplitude of the transducer waveform can be used to determine conveniently the physical distance to the break-off point. Increasing the amplitude decreases the break-off distance.

Once accurately adjusted, the circuits controlling the transducer and droplet charging and the regulators controlling the sheath pressure seldom cause instability in the droplet formation, assuming they are set to the optimum values. Clogging of the flow chamber walls and orifice by clumps of cells or debris, however partial, and the presence of air bubbles will disturb the smooth flow of the sheath fluid and cause instability in droplet formation. The temperature of the sheath fluid is another important factor, as changes in temperature will alter the viscosity of the sheath fluid and with it the speed of the jet. A 2 °C change in temperature will produce a 7.5% change in viscosity and an 11.5% change in the distance to the break-off point. If the room temperature changes during sorting (it usually increases due to heat generated from transformers and electronic circuits), the break-off point and thus the drop delay will also drift and sorting yield and ultimately purity will be affected. For this reason it is recommended that air conditioning is installed to maintain a constant temperature in the room containing the sorter.

3.2 Anti-coincidence and sort-mode settings

Because the arrival of particles at the laser intersection is a stochastic process, the possibility exists, particularly at high sample-event rates, that cells may be so close together that those not fulfilling the sorting criteria may be sorted coincidentally with selected cells. It is clear that such events not only affect the accuracy of the sort counting but, more significantly, will also compromise sorting purity. Anti-coincidence gating enables sorting purity to be maintained in these circumstances. When anti-coincidence gating is applied, the system determines whether an event has occurred in the time relating to those droplets which will form before and after the charged droplet. If a second event is flagged within this time window, the charge pulse is cancelled and the selected cell, together with the possible contaminants, goes to waste. While anti-coincidence gating preserves the purity of the sort, the cell yield is reduced due to the aborted sorting events. When selected cells are very rare, it may be more important to isolate every selected cell irrespective of possible contamination so that no selected cells are lost. In this case the anti-coincidence gating is turned off, a setting often known as 'enrich mode'.

A second consideration is the accuracy of droplet charging. Although droplet formation is a stable process, it is inevitable that because of minor transient disturbances of the flow and more gradual drifting of the break-off point due to changes in temperature or sheath pressure, the charge pulse timing may not coincide with the intended droplet formation. In this case an incorrect droplet will be charged and the selected cell will not be sorted, hence reducing the sorting yield. To overcome this problem, it is common to charge more than one droplet (usually two or three) for each sorting event, thus increasing the chance of isolating the selected cell. Inevitably, sorting more than one droplet will increase the number of coincident events, thus lowering the yield when anti-coincident gating is on or decreasing the purity in enrich mode.

Most instruments allow the additional refinement of statistical droplet deflection. Here, the system determines whether a selected cell event is in the centre or outside quarters of the time window defining the future droplet formation. If the event is not in the centre half of the time window, the system charges an additional droplet (e.g. two droplets rather than one) to selectively increase the chance of sorting the particular cell. A similar sophistication can be applied to the anti-coincidence gating when absolute accuracy of sorting counts is required, as in the case of single particle sorting. Here, the same time window relating to the future droplet formation is examined, but if the selected cell is not within the centre half of the window the sort is aborted. This sorting mode, which can be known as 'phase gating', ensures that precisely the desired number of cells are sorted when sorting yield is unimportant.

4 Practical considerations

4.1 General comments

Several basic factors must be considered when preparing a flow cytometer for sorting. The instrument should well maintained and cleaned as, in particular, dried saline deposits around the nozzle assembly can affect the transducer waveform and charging pulses. The sheath fluid should be particle-free as blockages of the nozzle are the most frequent cause of sorting problems. Similarly, care must be taken in sample preparation to ensure that clumps and debris are not present to cause blockages. Sample concentrations are usually higher than for analysis so as to enable a high frequency of events at small sample core diameters. Typically, these are between 2 and 6×10^6 cells/ml. For sorting viable cells, which often takes several hours, it is clearly necessary to maintain cell viability by using an appropriately buffered medium and, when necessary, cooling the sample and collection tubes.

The general steps required for operating a droplet sorter are set out in *Protocol 1*.

Protocol 1

Operation of a droplet sorter

Method

1. Establish stable normal sheath flow and optical alignment according to routine practice.
2. Adjust the sheath pressure according to the nozzle orifice diameter as recommended by the instrument manufacturer.
3. Switch on the transducer drive and adjust the transducer amplitude to about three-quarters of maximum. Where available, adjust the transducer frequency while observing droplet formation until the minimum break-off distance is achieved while minimizing the number of satellite droplets visible (usually two).
4. Turn on the high voltage to the sorting plates and, where available, adjust the centre position of the stream of uncharged droplets. Switch on the test pulse and adjust the phase angle (charge control angle) until stable, sharply defined, single

deflected streams are achieved. It is normal for two, close centre streams to be visible.

5. Set the break-off distance according to the manufacturer's recommended procedure. This will vary from instrument to instrument, but check it regularly using the procedure described in *Protocol 2*.

6. Select the number of droplets to be sorted and the anti-coincidence sorting mode

7. To verify correct sorting settings, run standard particles or cells visible by light microscopy and set appropriate sorting gates. Sort a specific number of particles (e.g. 50) directly on to a clean microscope slide and count the actual number of sorted particles using the microscope. A close agreement between the number of particles sorted as reported by the instrument and as counted directly indicates correctly adjusted sorting conditions.

4.2 Setting the sorting delay

The sorting delay is usually set in units of transducer wavelengths. With 'stream in air' systems, the distance from the laser intersection point to the break-off point can be measured and divided by the distance between droplets (the distance between 10 droplets divided by 10). This sorting delay can then be entered into the sorting logic control. In many instruments, this procedure can be performed semi-automatically.

If the sorting delay can only be set as an integer, and the calculated sort delay is not a whole integer, the transducer amplitude can be adjusted to alter the physical distance to the break-off point, the phase angle adjusted to re-establish single stable side-streams, and the drop-delay measurements remade. This process is repeated until a sort delay of a whole integer is attained. Although measurements of droplet wavelength and break-off distance allow direct calculation of the drop delay in these systems, it is still recommended that test sorts as described in *Protocol 2* are used routinely to confirm correct sorting operation.

Protocol 2

Empirical determination of sort delay

Method

1. Establish correct sorting conditions as described in *Protocol 1* and define sorting gates for cells or test particles.

2. Set the sort delay to a value lower than expected and sort a specific number of cells or particles on to a microscope slide. Increment the sort delay, move the microscope slide, and sort the same number of particles to the second position. Repeat this procedure until five sort delays have been tested.

Protocol 2 continued

3. Using appropriate microscopy (phase or fluorescence) count the number of particles in each sort position. Identify the most appropriate sorting delay by the position containing the greatest number of particles.

4a. If the number counted in the best position is close to 100% of that expected, proceed to sort using this sort-delay setting.

4b. If particle counts are distributed over two positions, the correct sorting delay is between the delays corresponding to these two positions. If fractions of transducer wavelengths can be set, repeat the test sort pattern using smaller sort-delay steps to identify the best sort delay. If only whole integer delays can be set, turn on the test sort mode and make a small adjustment of the transducer amplitude followed by adjustment of the charge angle to re-establish single stable side-streams and repeat the test sort pattern. Repeat this process until close to 100% of the particles sorted can be counted in a single drop position, thus identifying the correct sort-delay setting.

5. If sorted particles are detected in more than two of the sorting test positions, un-stable droplet formation or disturbed flow is indicated. Clean the nozzle and check for the absence of debris or air bubbles, then repeat the sorting set-up procedure.

4.3 Determining sorting yield and purity

If sufficient cells are sorted, the absolute yield of cells is again best determined practically, as losses can easily occur which can make the number of sorted cells reported by the instrument inaccurate.

Protocol 3

Estimating sorting yield

Method

1. Before sorting, measure the starting volume of the cells and estimate the concentration using a haemocytometer or Coulter counter. Take great care in measuring the cell concentration as this is a common source of error in the determination of yield. If used incorrectly, the haemocytometer is notoriously inaccurate for esti-mating cell concentration.

2. Determine the percentage of cells within the sorting gates from the instrument display.

3. After sorting, measure the volume of unsorted sample remaining. Measure the vol-ume of the sorted fraction and estimate the concentration using a haemocytometer or Coulter counter.

4. Analyse a sample of the sorted cells to determine the percentage of cells within the original sorting gates (sorting purity). Note that many cells can remain within the sample tubing after routine backflushing. If the sorted sample in which cells are at a low concentration is analysed immediately, an underestimate of purity can often result due to contamination by these retained cells. Avoid this problem by running a sample of dilute detergent for a few minutes followed by clean saline until events due to retained cells are acceptably infrequent. Note that the fluorescence intensity of the sorted fraction is also often slightly less than the original gated population due to loss of stain during the sorting process or bleaching of the fluorochrome during exposure to the laser.

5. Multiply the volume of the original sample used by the concentration to calculate the total number of cells processed. Multiply this by the proportion of cells gated to give the expected number of cells sorted.

6. Multiply the volume of the sorted fraction by the concentration to calculate the total cells sorted. Multiply this by the proportion of sorted cells within the sorting gates to give the observed number of cells sorted.

7. Express the observed number of cells as a percentage of the expected number of cells to give the absolute sorting yield.

8. Calculate the instrument yield by expressing the observed number of cells sorted as a percentage of the number reported by the instrument statistics. The instrument yield takes into account aborted sort events due to anti-coincidence gating and gives a measure of the accuracy of the instrument settings and the efficiency of cell collection.

Many parameters will affect the yield. Inaccurate setting of the droplet delay, instability in droplet formation, and movement of the break-off point will drastically reduce the yield. In such situations, droplets before or after the droplet containing the selected cell will be deflected in error. At best, these droplets will not contain cells and purity will not be affected. However, it is a simple matter to monitor the correct setting of the droplet delay if particles large enough to be seen under the microscope are being sorted. After setting the controls according to the normal procedure, a short sort of about 50 cells is performed on to a microscope slide. The drop of fluid is then observed under the phase-contrast microscope and the number of cells present quickly counted. A discrepancy between the number of cells sorted, as reported by the instrument, and the actual number counted is a direct indication of inefficient operation and will indicate the potential reduction in yield. The same procedure can be used during a long sort to monitor efficient sorting. Alternatively, the position of the break-off point can be observed through a microscope using stroboscopic illumination. Drifting of the break-off point can be detected by reference to an eyepiece graticule and appropriate action taken (adjustment of transducer amplitude).

The accuracy needed for setting the deflection delay increases with the size of the particle being sorted. As the size of the cell increases, its physical passage through the orifice will cause minor perturbations in droplet formation and uncertainty in the arrival time of the cell at the break-off point. One approach to reduce this affect is to deflect more than one droplet per cell so maximizing the chance of successful sorting. Alternatively, a larger orifice can be used. However, the lower transducer frequency required for a larger orifice will reduce the absolute sorting rate of the system.

4.4 Collection vessels

To collect large numbers of cells, they are usually sorted into centrifuge tubes. If plastic tubes are used, during a long sort, charge can build up on the tubes to such an extent that charged droplets may be deflected, thus reducing the collection efficiency. If this problem arises, an earthed platinum wire should be inserted in the collection tubes. Static electricity causes fewer problems with glass collection tubes because the charge is dissipated more efficiently across the glass surface. Collection tubes can be coated with protein (e.g. with fetal bovine serum) before sorting to discourage cells from adhering to the walls of the tubes. It is difficult to prevent fixed cells from sticking to the tube.

If cells are required for amplification of their DNA by a polymerase chain reaction (PCR), they can be sorted directly into the PCR reaction tube.

If cells, which grow attached to plastic surfaces, are being sorted for culture, they can be sorted directly into the wells of a multi-well culture plate, which should be about one-third full of culture medium. After the sort, incubate the cells at 37°C for 1–2 h to allow the cells to settle and attach to the plastic surface, then carefully remove three-quarters of the supernatant and add fresh medium.

Cells sorted in small numbers for identification can be collected directly on to a microscope slide. They can also be sorted into small wells on specially manufactured slides which are covered with PTFE film with circular holes cut in the film. Cells will adhere more easily to the glass if the slides are pre-coated with polylysine. After giving the cells time to settle, the slides can be immersed gently into a fixative and the cells stained for microscopic examination.

If the desired cells are in a low concentration, the rate of sorting on to a microscope slide will be low and the sheath fluid will evaporate during the sort. The salt will crystallize and the sorted cells will shrink and lose their morphology. The problem can be reduced by using a lower salt content in the sheath fluid (diluted 1 in 50).

If the cells are being sorted to use as dot blots in DNA hybridization, they can be collected directly on to a nylon filter. It is not difficult to construct a filter holder mounted in a small chamber to which a slight vacuum is applied, this allows the sheath fluid to be removed as the droplets land.

4.5 Sterile sorting

For many applications it is necessary for the sorted fractions to be prepared aseptically. In most cases, the instrument can be effectively sterilized by running 70% ethanol as the sheath fluid and as a sample for between 20 and 30 minutes. During this period, the surfaces of the instrument, particularly around the sort-collection and flow-chamber areas, are swabbed with 70% ethanol. Using good sterile procedures, a sterilized sheath bottle containing sterilized sheath fluid (usually PBS) is attached to the sheath lines. A sterile 0.2 micrometre filter can be inserted between the sheath container and the sheath line to provide additional protection against contamination. Similarly, sterile sheath fluid in a sterile tube is run as a sample to flush out the 70% ethanol. The instrument is allowed to stabilize for between 30 minutes and 1 hour and then adjusted for sorting. As the test beads normally used for routine optical alignment and sorting adjustment cannot easily be sterilized, 70% ethanol followed by several changes of sterile sheath fluid are run as samples to re-sterilize the sample lines.

5 Specialized features

5.1 Sorting single cells

Some commercial cell sorters have a special holder for sorting cells into the wells of tissue culture plates. The number of cells to be sorted into each well can be specified and, when the specified number has been delivered, the holder moves to position the next well for receipt of cells. A common application is for cloning a selected population; single cells are sorted into the wells of a 96-well plate. Single droplet deflection should be used with 'phase gating' (see Section 3.2 above).

5.2 Sorting rare events

Rare events present a particular problem because of the time needed to collect enough cells. The sample can be enriched by pre-sorting at a high flow rate in 'enrich mode' (giving high yield but low purity) and then re-sorting for high purity.

An alternative method of pre-enrichment is to remove a high percentage of unwanted cells by magnetic separation (1). This strategy can only be adopted if there is an antigen expressed on the surface of 90% or more of the unwanted cells which is not expressed on the rare cells.

5.3 High-speed sorting

Sorting large numbers of cells or rare event sorting is enhanced if the rate of throughput of the cells can be increased. The rate of sorting is a trade off between the purity required and yield. In conventional cell sorters, the maximum flow rate lies between 7000 and 15 000 cells/sec, depending on the configuration of the flow cell. The sorting rate can be increased by raising the droplet frequency, which necessitates an increased sheath flow pressure and therefore a re-design of the flow system to withstand the higher pressures needed.

Beckman–Coulter and Becton Dickinson both supply sorters which will run at flow rates up to 25 000 cells/sec (Epics Altra Hypersort and the FACSVantage SE TurboSORT, respectively). Cytomation have produced an instrument, specifically designed for high-speed sorting (rather than a modification of an existing instrument), (MoFlo MLS) which will run at speeds in excess of 25 000 cells/sec.

6 Other flow-sorting systems

While electrostatic droplet deflection is the most widespread sorting method, other systems are available on commercial instruments. At the time of writing, these are the fluidic switch system of the PAS instruments (Partec) and the catcher tube mechanism of the FACSCalibur (Becton Dickinson).

The PAS sorters utilize the microscope-based flow chamber shown in *Figure 3a* of Chapter 1, modified so that the channel downstream of the detection point bifurcates to form a 'Y'-shaped channel. The 'Y' channel is constructed so that two-thirds of the flow passes down one channel (waste arm) and one-third passes down the other channel (sorting arm). Positioned within the waste arm is a piezoelectrically actuated piston, which when operated causes a pressure pulse to pass back up the waste arm temporarily diverting the majority of the flow into the sorting arm. Unsorted cells pass naturally into the waste arm with the two-thirds of the flow. When a cell falls within the sorting criteria, the piston is actuated after the appropriate delay and the majority of the flow is temporarily diverted into the sorting arm and with it the selected cell. Sorting rates in the order of 1000 cells/sec can be achieved with this system

The FACSCalibur utilizes a piezoelectrically operated 'catcher' tube which, in its non-sorting mode, is offset from the centre of the downstream side of an enclosed flow chamber (see *Figure 3*). Sheath fluid passes continually through

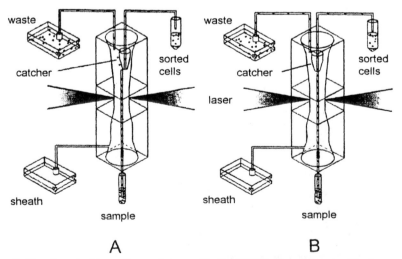

A   **B**

Figure 3 The piezoelectric sorting system used in the FACSCalibur cell sorter (Becton Dickinson). Note that the direction of flow is vertical. (A) The catcher tube is at the edge of the sheath stream; unwanted cells are deflected to waste. (B) The catcher tube has been moved to the centre of the sheath stream; the desired cells are collected.

the flow chamber and into the 'catcher' tube, but, as the normal cell trajectory is coaxial due to hydrodynamic focusing, unsorted cells pass to the side of the 'catcher' tube and do not enter it. When a cell is detected which fulfils the sorting criteria, the 'catcher' tube is actuated and moves into the centre of the flow to intercept the selected cell, subsequently returning to its offset position. In this way up to 300 cells/sec can be sorted.

Neither of these systems offers the higher sorting rates achieved with droplet sorting, and sorting is restricted to the separation of a single subpopulation. However, a major advantage is that these sorting systems are enclosed thus avoiding aerosol formation and so reducing the possible risk from biohazardous samples.

Further reading

Ormerod, M. G. (1998). In *Cell separation: a practical approach* (ed. D. Fisher), p. 169. IRL, Oxford.

Pinkel, D. and Stovel, R. (1985). In *Flow cytometry instrumentation and data analysis* (ed. M. A. Van Dilla, P. N. Dean, O. D. Laerum, and M. R. Melamed), p. 77. Academic Press, Orlando, FL.

Lindmo, T., Peters, D. C., and Sweet, R. G. (1990). In *Flow cytometry and sorting* (2nd edn) (ed. M. R. Melamed, P. F. Mullaney, and M. L. Mendelsohn), p. 145. Wiley–Liss, New York.

Reference

1. Dyer, P. A., Brown, P., and Edward, R. (1998). In *Cell separation: a practical approach* (ed. D. Fisher), p. 191. IRL, Oxford.

Chapter 5

Immunofluorescence of surface markers

Michael R. Loken, Cherie L. Green, and
Denise A. Wells
HematoLogics Inc., c/o Fred Hutchinson Cancer Research Center, Seattle, WA, USA

1 Introduction (1, 2)

The discovery of monoclonal antibodies by Köhler and Milstein in 1975 dramatically enhanced the use of immunofluorescence for identifying cell-surface antigens. When applied in the flow cytometric analysis of heterogeneous populations, these highly specific monoclonal antibodies have broadened our understanding of diseases such as the progression of HIV infection, the origin and nature of leukaemia and lymphoma, and the regulation of haematopoietic cell differentiation and maturation.

The discovery of new fluorescent chromophores has further exploited the power of quantitative cellular analysis. These new dyes can be used in combination with the standard fluorescein isothiocyanate (FITC) to examine multiple antigens simultaneously. The orange and red emitting dyes—phycoerythrin (PE), tandem conjugates, and peridinin–chlorophyll (PerCP)—can all be excited by a single laser wavelength with their emission separated by bandpass filters (see Chapter 2).

The information provided by the additional parameters combining light scatter and two or more immunofluorescence colours is crucial for the enumeration of cell types in heterogeneous mixtures of cells. However, the data become far more complex than simple, single-colour analysis of homogeneous cell populations.

The goal of flow cytometric immunofluorescence analysis is to use the light-scattering and antigenic characteristics of each cell in sequence to assign that cell to a specific group of cells having similar properties. This primary goal of population assignment must be kept in mind when performing surface marker analysis and in developing quality control procedures and rules for data analysis. Objective criteria must be used to determine whether a cell more closely resembles population A or population B. This issue of population assignment is the basis for determining whether or not to move the fluorescence cursors and/or for defining the proper negative control for a specific specimen. Quality

61

control procedures must be instituted to make these distinctions reproducible, so that data can be compared from one time point to another and from one laboratory to the next. These quality control procedures must include: specimen collection and transport; sample processing; instrument standardization; data collection and analysis; as well as procedures for troubleshooting difficult cases.

The first step in immunofluorescence analysis is to identify the cells of interest. For lymphocyte subset analysis this requires setting a lymphocyte light-scattering gate. The lineage assignment of leukaemic cells using immuno-fluorescence requires leukaemic cells to be distinguished from their normal counterparts. This first stage of immunofluorescence analysis is simplified when analysing cell lines as they are homogeneous rather than heterogeneous.

Once the cells of interest are distinguished from other cell types, immuno-fluorescence can be used to determine the proportion of that cell type relative to other cells in the heterogeneous population. Lymphocyte analysis for monitoring the progression of human immunodeficiency virus (HIV) disease, for example, concerns cell-subset proportions. Variations in processing or analytical procedures can affect the proportion of cells identified in the results and are central issues in quality control.

Immunofluorescence can also address changes in the intensity of antigen expression of a cell population relative to a disease process or external experi-mental stimulus. In lineage assignment of leukaemia, for example, the propor-tion of leukaemic cells is less important than the detection and intensity of specific antigens on their surfaces. Quality control in this setting focuses more on measurements of antigen intensity than on cell recovery.

Other techniques can be combined with immunofluorescence to identify cellular characteristics of the selected population. The DNA ploidy analysis of a specific cell population identified by immunological criteria is an example of combined techniques. Quality control in this case requires that subsets of the cells of interest are not selectively lost during processing or analysis. Addition-ally, immunofluorescence and other techniques must not interfere with cell identification and analysis required by the data.

Certain immunofluorescence techniques may involve a combination of these basic applications. For example, the enumeration of activated lymphocytes re-quires the determination of both the proportions of activated cells and the intensity of expression of the activation antigens. The basic principles of quanti-tative immunofluorescence must be applied to assure quality control of the analyses whether performing one-colour or three-colour immunofluorescence.

With respect to quality control, the emphasis may vary depending on the type of question being asked. However, certain general principles must be addressed in a comprehensive quality control programme to ensure reproducible, accurate results. This chapter will focus on the use of monoclonal antibodies in the measurement of cellular subsets, emphasizing the features of quality control that provide consistent, accurate results. Although the chapter primarily deals with immunofluorescence of surface markers on human peripheral blood lymphocytes, the general principles will apply to other situations.

2 Instrument standardization

Instrument standardization is performed at the beginning of each run to confirm that the instrument is operating under standard parameters and that it meets minimum performance requirements. The optics (including light-scatter detectors and fluorescence detectors) and electronics (including photomultiplier tube voltages and spectral compensation), as well as expected fluorescence staining patterns of the monoclonal antibody reagents, must be tested daily.

2.1 Fluorescence standards

Several requirements for flow cytometric standards are recognized. The standard must be able to test all optical channels including forward light scatter, right-angle light scatter, and all fluorescence channels, and must be stable so that it can be stored for a long period. It may be a single particle with a broad-spectrum emission or a series of particles, each testing a single channel.

The standard must have well-defined peaks in each of the channels being tested. The signal intensity of the standard must be similar to that obtained from the cells or particles being tested. It is improper to set up the instrument for immunophenotyping using a standard that is much brighter, much larger, or much smaller than lymphocytes.

Standards are usually refrigerated as a concentrate. A working suspension is then made daily for use only on that day. Often, the diluted standard may change its characteristics after 24 h in diluent. The standards should be used at intervals during the day to check the instrument sensitivity continuously. This requires that the standard is easy to obtain and relatively inexpensive so that there is little concern about frequent use.

2.2 Optical alignment

Optical alignment for many flow cytometers is fixed and may only be adjusted by company service representatives. For those instruments in which the alignment can be changed by the operator, a standard procedure must be followed. Standard particles are used to adjust the alignment to ensure that the cells are flowing through the intersection of the laser beam and the optics that collect the signals. This procedure must also be performed to make sure that each detector is optimized to give maximal and reproducible signals from standard particles. The goal of optical alignment is to optimize the signal in each of the sensors. Once achieved, the electronics must be set for routine operation.

2.3 Electronic standardization

There are two approaches to electronic standardization: one method requires using exactly the same electronic settings from one day to the next and monitoring where the standards appear in each of the five (or more) channels. An alternative method involves adjusting the electronics to place the standards in exactly the same channels every time, thereby recording the minor changes in

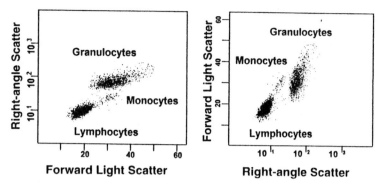

Figure 1 Forward and right-angle light-scatter signals are used for positioning the cell populations on scale. Light-scatter cytograms of leucocytes from human peripheral blood (compare this figure with *Figure 13*, Chapter 1, which shows a linear display of right-angle light scatter).

electronic settings. These adjustments are usually small compared to the high voltage applied to the photomultiplier (PMT) on the fluorescence detectors. Similarly, minor adjustments can be made on the gain settings for the linear amplified channels (forward and/or right-angle light scatter). It is important to decide which approach will be used by the laboratory, to use it consistently, and to use it on a daily basis. A logarithmic scale (or non-linear scale) is often used for right-angle light scatter since there is a large dynamic range between the lowest and highest scattering cells (3).

The following procedure should be followed the first time the instrument is standardized using broad-spectrum standards (full-spectrum beads emit light into all fluorescence channels). The standardization assumes that the proper alignment procedure has been accomplished and the optical paths have been optimized.

Protocol 1

Initial standardization of the instrument

Caution: All biological safety procedures must be followed when handling human blood products.

Method

1. After the alignment procedure has optimized the optics, prepare a cell suspension of leucocytes from whole blood by NH_4Cl lysis of erythrocytes (see *Protocol 4*). Fix the cell suspension in 1% buffered paraformaldehyde (pH 7.2–7.4, see *Protocol 4*) in saline and dilute to 10^6 cells/ml.

2. Run the cell suspension through the flow cytometer, observing the forward and right-angle light-scatter dot plots (linear/log or linear/linear). Adjust the gain setting on the forward light-scatter amplifier to set the centre of the lymphocyte cluster to

Protocol 1 continued

just below the centre of the forward light-scatter axis (see *Figure 5.1*). Similarly, adjust the linear gain (or photomultiplier tube (PMT) voltage for log amplification) on the right-angle light scatter until granulocytes are displayed on the screen. Record the gain settings for the forward and right-angle light-scatter detectors in the notebook to be kept with the flow cytometer.

3. With unstained cells flowing through the instrument, display FL1 vs. FL2 dot plots (log, log). Increase (or decrease) the high voltage on the FL1 PMT so that >90% of the signals from the unstained cells appear in the first decade of fluorescence. Adjust the high voltage for the FL2 PMT until the unstained cells are in the first decade for that parameter. Repeat this procedure for all other channels (FL3, FL4, etc.). Record the settings of the high voltages for all the parameters (forward light scatter, right-angle light scatter, FL1, FL2, FL3).

4. Run the broad-spectrum standard, diluted one drop in 1 ml saline. Record the positions of the peaks in all the parameters.

5. Repeat steps 1–4 at least 10 times with different normal blood specimens. This will provide a measure of the variability of the electronic set-up for the 10 different normal donors.

6. Determine the average position of the peaks in each of the parameters for the 10 different donors. Use this average position of the broad-spectrum standard for subsequent set-up of the electronics of the instrument.

A routine procedure should be used to set up the electronics of the flow cytometer at the beginning of each run. The standards should be placed at the same intensity position each time and the electronics settings recorded (or, the electronics settings should be established and the peak positions of the standards should be monitored). As part of this standardization protocol, an unstained, fixed blood sample should also be analysed and the peak positions recorded. This provides a record of the instrument performance by which to assess its operation on a longitudinal basis. The procedure is essentially the reverse process used in the initial standardization.

Protocol 2

Daily standardization using broad-spectrum standards

Method

1. Align the instrument as outlined above. As a starting point, set the electronics settings to the same levels used on the previous day.

2. Make a dilution of one drop of broad-spectrum standard in 1 ml saline, fresh every day. Run the standard with the displays set to forward vs. right-angle light scatter (linear, log) and FL1 vs. FL2 (log, log). Place the peaks of the standard in the

Protocol 2 continued

positions determined in the initial standardization by adjusting the gain settings for the forward light-scatter parameter and the high-voltage settings on the right-angle light-scatter and fluorescence parameters. On a stable instrument, the settings should be minimally different from those used the previous day to place the standard in the same position. Record the daily settings in the flow cytometer log book.

3. Prepare a sample of unstained, fixed, normal blood leucocytes using the laboratory standard protocol. Dilute these cells to 10^6 cells/ml and run them through the flow cytometer. Record the daily positions of the lymphocytes, monocytes, and granulocytes with respect to the light-scatter parameters in the flow cytometer's log notebook.

4. Gate the fluorescence data based on the lymphocyte population. Record the daily medians of the fluorescence distributions for the unstained lymphocytes in each of the fluorescence channels in the flow cytometer's log notebook.

5. Plot the light-scatter and fluorescence detector settings and normal blood peak positions in a Levey–Jennings display (see *Figure 2*).

6. Plot a graph of the median fluorescence of the unstained lymphocytes and the position of the broad-spectrum standard beads in each of the channels.

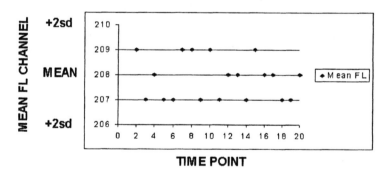

Figure 2 Example of a Levey–Jennings plot.

The graph of the electronic settings and peak positions provides a way of monitoring instrument stability over time. These graphs should be relatively straight lines if the instrument is stable. Misalignment or problems with the optics or electronics will be easily noted by comparing the settings required to place the standard in the same position as determined in earlier runs. A steady decline in one of the detectors will be evident. A rapid change resulting from PMT damage or misalignment can also be detected by a change in all of the detector settings.

The second graph measures the sensitivity of the instrument. This line will remain level if the sensitivity does not change over time. If the instrument becomes less sensitive, the median fluorescence of the unstained cells will increase relative to the position of the standard. A reduction in sensitivity will also be reflected in an increase in the voltage required to place the standard in the identical position, as previously observed.

2.4 Spectral compensation

When no compensation is set, cells labelled only with PE can be detected not only in the orange (PE) detector but also in the green (FTIC) and red (tandem or PerCP) channels. This overlap of spectral emission can be corrected electronically (see Chapter 1, Section 4.3.4). Tandem conjugates of PE and an acceptor dye, such as Texas Red or PC5, also have to be corrected as the emission of the dye into the orange (PE) detector depends on the efficiency of energy transfer from PE to the acceptor molecule. Each tandem reagent must have the compensation set independently, because quenching of the donor, PE, may vary from lot to lot so that the effective spectrum of the dye changes with different batches of reagent (4). The spectral compensation must be reset if the optical filters or the PMT voltages are changed on the fluorescence detectors.

When the compensation network is properly adjusted, cells labelled with FITC only record the same fluorescence as unlabelled cells when observed in the orange or the red channel. Cells labelled with phycoerythrin only record the same fluorescence as unlabelled cells when observed in the green and red channels. Proper setting of the compensation network is best demonstrated by examining results such as those shown in *Figure 3*. The data represent a mixture of cells that were unstained, stained with CD8–FITC only, and stained with CD4–PE only.

The populations of singly labelled cells should form right angles when displayed as a two-parameter dot plot. Undercompensation occurs when signals from the singly labelled cells are being detected in the other channels. Setting the compensation network too high (overcompensation) places the singly labelled cells off-scale on the low end in a correlated dot plot. The mean channels of the positive cells are then lower than the unstained cells when observed from the opposite channel.

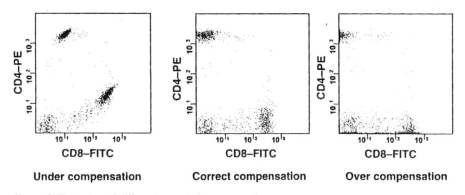

Under compensation   **Correct compensation**   **Over compensation**

Figure 3 Examples of different spectral compensation.

Protocol 3

Setting spectral compensation

Method

1. Prepare four groups of stained cells (see *Protocol 4*):
   - unstained or unlabelled cells;
   - fluorescein-stained cells (e.g. CD3–FITC or CD8–FITC);
   - phycoerythrin-stained cells (e.g. CD19–PE or CD8–PE);
   - PerCP-stained cells (e.g. CD3–PerCP or CD45–PerCP).

2. Use the same antibodies as those to be used in the subsequent tests to obtain the best results. The stained cells should have the brightest expected signals. Note that the best antibody for setting spectral compensation is one that displays a broad range of antigen-intensity distribution such as CD8. The CD8 antibody stains cells both brightly and dimly over the entire range of intensities, which permits proper spectral compensation.

3. The unstained cells should reside in the lower left-hand corner on a fluorescence display of phycoerythrin vs. fluorescein (see *Protocol 1* and *Figure 3*). Set FITC-labelled cells so that they are detected only in the fluorescein channel and appear identical to unstained cells when observed in the phycoerythrin channel. Similarly, set the PE-labelled cells to be detected only in the phycoerythrin channel and to appear as unstained in the fluorescein channel.

4. Repeat step 3 using unstained cells and CD8–PE and CD8–PerCP stained cells. The unstained cells should sit in the lower left-hand corner on the fluorescence display of phycoerythrin vs. PerCP (see *Protocol 1*). Set the PE-labelled cells so that they are detected only in the PE channel and appear identical to unstained cells when observed in the PerCP channel. Test cells stained only with FITC and PerCP for any spectral compensation between these two channels. Compensation for tandem conjugates can also be set using these procedures, substituting these dyes for PerCP.

3 Testing sensitivity and resolution

In order to assess instrument sensitivity, it is necessary to have at least two intensity standards, measuring the separation between them. If the instrument is stable (i.e. sensitivity remains constant) the relative positions of the two standard particles will remain constant. If the sensitivity is reduced, the distance between the two standards will decrease.

It is possible to use a single standard as one of the reference points and unstained normal cells as the second reference point. The unstained cells, however, may vary from day to day so that a running average must be used.

It is important to show that the instrument is sensitive enough to perform the test. The instrument is adequately sensitive if it can distinguish between the autofluorescence of unstained cells and a particle that is dimmer than unstained cells, such as erythrocytes or a certified blank (see *Figure* 4). If the instrument can detect dimmer signals than unstained cells, it can identify dim but positively stained cells. This demonstrates that the sensitivity of the instrument is limited by the sample being analysed. Unstained cells have intrinsic fluorescence that comes from the biological components within the cell. This fluorescence, called autofluorescence, varies with cell type. For example, lymphocytes exhibit the least autofluorescence of white blood cells, whereas monocytes and granulocytes have the most. The autofluorescence of unstained lymphocytes is equivalent to approximately 1000 fluorescein molecules per cell. Therefore, the lowest limit of detection of the flow cytometer is controlled by the sample itself and not by the instrument. Sensitivity can also be affected by the flow rate of cells through the laser beam. The longer the cell is in the laser beam, the more fluorescence it will emit, thereby increasing the amount of fluorescence coming from a given number of fluorochromes. If the sheath pressure or flow rate is changed on the instrument, the standardization procedure must be repeated.

If the instrument does not meet the required specifications, it is not known *a priori* whether the difficulty is within the instrument or if the standard has changed. One method of testing the standard to ensure that it has not changed its characteristics is to have it cross-calibrated with a secondary standard.

If both the primary and secondary standards change positions, it can be concluded that the instrument sensitivity has changed. In a similar manner, the light-scatter channels must be standardized and the positions of the standard particles should be related to standard cell preparations to ensure that discrimination between cell types by light scatter is constant.

The ability to distinguish cell populations lies with the optical adjustment of the flow cytometer and the quality of the cell preparation. Different sample preparation techniques can cause cells to swell or to shrink, and fixation can cause changes in the refractive index of the cells. Therefore, changing sample processing protocols can affect the resolution observed on the same instrument. To minimize variation and maximize consistency, standardized sample processing procedures should be used. A progressive or dramatic change in light-

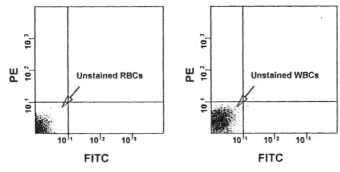

Figure 4 Instrument sensitivity can be evaluated by comparing the low-level signals of erythrocytes (RBCs) with the autofluorescence of unstained lymphocytes (WBCs).

69

scattering resolution on a single instrument suggests a change in the optical characteristics. This may require realignment of the instrument or a service call by a trained representative.

4 Immunofluorescence

The two most commonly used fluorescent dyes in flow cytometry are fluorescein (green) and phycoerythrin (orange). As a third dye, peridinin–chlorophyll-*a* (PerCP) is commonly used (see Chapter 2, Section 3). Although PerCP can be excited at 488 nm, this molecule has a large Stoke's shift with an emission maximum in the deep red at 690 nm.

The combination of antibody conjugates of FITC, PE, and PerCP can be used for the simultaneous determination of three different antigens on the cell surface. Texas Red–PE (ECD) or Cy5–PE (Cy-chrome or PC5) conjugates may also be used to supply the third colour. If a He–Ne laser is fitted to the flow cytometer, allophycocyanin may be used to supply a fourth colour (see Chapter 2, Section 3).

5 Staining for surface markers (5)

The detection of multiple cell-surface markers is important in the precise identification of lymphoid cell subsets as well as malignant proliferation of undifferentiated myeloid and lymphoid cells. The simple one-colour approach may not specify the cell type of interest, thus two or more cell-surface or cytoplasmic antigens should be measured. Non-specific binding can affect cell discrimination when the antibodies attach to cells through mechanisms other than the specific binding site. Such non-specific binding is increased when antibody aggregates form during the manufacture or storage of the reagent. The Fc receptors on cells (particularly monocytes) bind different classes of immunoglobulins and can be confused with specific monoclonal antibody staining. Charge differences between cells, particularly dead cells, can result in non-specific staining. Since non-specific binding can vary between different reagents, it is best to use a properly selected reagent panel to control for such non-specific reactivity. *Protocol 4* describes multicolour immunofluorescent staining procedures and can be modified for one- or two-colour staining.

The simultaneous measurement of four or five independent parameters (light scatter and fluorescence) on the cells passing through the laser beam creates a multi-dimensional space in which cell populations with dissimilar properties emerge in different locations as clusters of cells. One of the difficulties in multi-parameter flow cytometry is to display and to identify cell subpopulations in these multi-dimensional data files. The use of colour facilitates the identification of cell populations in different two-dimensional projections of the data. Populations of cells identified in one, two-dimensional projection of the four- or five-dimensional data can be depicted with a primary colour and appear as the same colour in other projections of that data.

Protocol 4

Procedure for three-colour immunofluorescent staining

Reagents[a]

- FITC-conjugated, PE-conjugated, and PerCP-conjugated antibodies
- Working PBS buffer: 1:10 dilution of $10 \times$ Dulbecco's PBS (Ca^{2+} and Mg^{2+} free) in distilled water. Stable for 2 weeks at $4\,°C$.
- PBS, 2% fetal bovine serum (FBS), 0.01% NaN_3 (monoclonal staining): 10 ml FBS and 0.5 ml of 10% NaN_3 in 500 ml of working PBS

- RPMI-1640 medium, 2% PBS, gentamicin
- NH_4Cl (erythrocyte lysing reagent): 8.26 g NH_4Cl, 1 g $KHCO_3$, 0.032 g Na_3EDTA in 1 litre of distilled water
- 1% paraformaldehyde. Dilute a 10% stock solution of paraformaldehyde 1:10 with working PBS.[b] Make up fresh each time it is to be used.

Method

1. If the assay is to be performed on tissue, isolate cells using gentle teasing, wash twice in RPMI-1640 medium, 2% PBS, gentamicin and resuspend at a cell concentration of $1.0 \times 10^7/ml$.

2. Label sample tubes with appropriate identification for the sample to be tested and with the combination of monoclonal antibodies to be used in that tube.

3. Add FITC-conjugated, PE-conjugated, and PerCP-conjugated antibodies in proper test amounts to the same tube for each three-colour test combination and dilute to 50 µl with PBS, 2% FBS, 0.1% NaN_3. Alternatively, make up cocktails containing the three antibody conjugates at proper concentration (times the number of tests to be used) in PBS (with 2% FBS and 0.1% NaN_3) to make each test volume equal to 50 µl. Note that a control tube will be necessary but will be set up only with mouse IgG–FITC and mouse IgG–PE.

4. Add one drop of peripheral blood or of a bone-marrow or cell suspension (50 µl) to the monoclonal antibody mixture in each tube. Agitate the tubes gently to mix and incubate for 20 min at room temperature in the dark.

5. Add 3 ml of warm ($37\,°C$) NH_4Cl erythrocyte lysing solution to each tube for peripheral blood or bone-marrow marrow specimens and mix by inverting. (If a cell suspension without contaminating erythrocytes is used, wash with the PBS, 2% FBS, 0.01% NaN_3.) Allow lysis to occur for 5 min at room temperature.

6. Spin for 5 min at 300 g. Aspirate the supernatant and agitate the cell pellet. Wash with 2 ml of the PBS, 2% FBS, 0.1% NaN_3.

7. Aspirate the supernatant and agitate the pellet. Resuspend the pellet in 0.5 ml of 1% paraformaldehyde.[b] Samples are ready for analysis after 30 min of fixation.

[a] The pH of the solutions must be checked every day. The pH should be between 7.1 and 7.4. If the pH is high, add Hepes buffer to bring it into the desired pH range.

[b] This solution is used to preserved the stained cells overnight for analysis the next day and to reduce biohazard materials in the specimens.

6 Titration of antibodies

Immunofluorescence reagents are usually titrated before they are used for routine testing of specimens to ensure proper quality control and to minimize lot-to-lot variation. Comparisons should also be made between lots when new antibodies are purchased. The labelling is performed using twofold (doubling) dilutions of test reagent and a standard number of cells ($\times 10^{6}$) in 100 µl of buffer for each test. The same volume of reagent should be used at each dilution point. The mean peak intensity for the target cells should be determined at each dilution and the fluorescence intensity plotted vs. dilution.

The correct concentration of reagent depends on the procedure used to process the cells. For procedures in which a wash step is used before fixation of the cells, the optimal dilution to choose is one dilution less than the maximum intensity plateau (see *Figure 5*). Background fluorescence for this optimal dilution should be no greater than the isotype control or unstained cells. For two or more colour reagents, a checker-board cross-titration should be used.

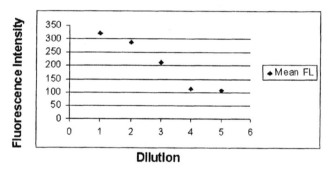

Figure 5 Optimal dilution for the titration of antibodies.

7 CD nomenclature

The CD (cluster of differentiation) nomenclature was developed to standardize monoclonal antibodies from different sources. Studies of antibody reactivity patterns for several cell types were analysed and compared with biochemical analysis to show similarities between them. The antibodies were then assigned to groups with similar characteristics and given a CD number. The antibodies listed in *Table 1* are commonly used in identifying various lymphocyte subpopulations.

8 Population assignment (6–8)

It should be kept in mind that flow cytometric analysis is based on assigning each cell to a group of cells with similar characteristics. Population assignment is facilitated when the groups of cells are completely isolated from each other and form discrete clusters (*Figure 6*). For example, the T-helper/inducer popula-

Table 1 Commonly used antibodies in lymphocyte subset analyses

CD45	Identification of leucocytes
CD14	Identification of monocytes
CD3	T lymphocytes
CD4	Helper/inducer T lymphocytes and monocytes
CD8	Cytotoxic/suppressor T lymphocytes and some NK cells
CD19	B lymphocytes
CD16	NK lymphocytes and neutrophils
CD56	Most NK cells and some T lymphocytes

tion, identified by CD3 and CD4, exhibits clear differences in cellular characteristics from other lymphoid and non-lymphoid cells and can be readily separated from the others. The assignment of a particular cell to its population becomes trivial when that population is discrete. One goal in selecting a reagent panel is to obtain discretely identifiable groups of cells.

Continuous populations occur when antigen expression for a group of cells is dim or not well resolved from the cell populations that do not express the antigen (see *Figure 6*). Continuous-staining populations are often seen with activation antigens such as HLA-DR, CD25, CD8, CD56, and CD38.

More care is needed in discriminating between positive cells and the negative population in this situation. Selection of the proper negative control becomes important. It may not be appropriate simply to use isotype-matched reagents. In addition, careful setting of compensation networks must be made for accurate and reproducible discrimination of the positive cells in a continuous population.

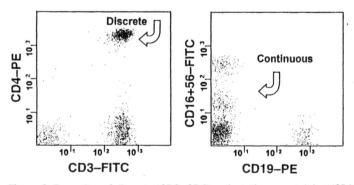

Figure 6 Examples of discrete (CD3, CD4) and continuous staining (CD19, CD16 + CD56).

9 Constructing an immunophenotyping panel (9, 10)

An appropriately selected immunophenotyping panel should provide the most accurate and unambiguous definition of populations and subpopulations. The reagents should be selected to maximize the information about an entire population, to provide an internal quality control check for each sample, and to allow

73

MICHAEL R. LOKEN *ET AL.*

Table 2 Recommended T-lymphocyte subset panel (10, 11)

Tube number	FITC	PE	Use
1	CD45	CD14	Light-scatter gating
2	CD3	CD19	Total T and B lymphocytes
3	CD3	CD4	Total T and CD4+ T lymphocytes
4	CD3	CD8	Total T and CD8+ T lymphocytes
5	CD3	CD16 and CD56	Total T and NK lymphocytes

consistency checks to be made between sample tubes. A useful panel must not only identify the population of interest, it must define a negative reference population by which to compare fluorescence intensities. A clearly conceived panel also defines internal consistencies and patterns. The T-lymphocyte subset panel recommended by the Centers for Disease Control in the USA (CDC) (10) is listed in *Table 2* and will be used as an example of a well-integrated reagent panel.

9.1 Identifying the cells of interest: CD45 and CD14 expression

CD45 and CD14 antigens have different expression patterns on white blood cells and can be used to establish the optimal light-scatter gate for lymphocytes or other cells:

(a) All leucocytes express the CD45 antigen, the common portion of the T200 antigen. Lymphocytes express high levels of CD45; monocytes express intermediate levels of CD45, whereas granulocytes express lower levels of CD45; erythrocytes, platelets, and non-haematopoietic cells do not express CD45.

(b) Since all leucocytes express CD45, it is not appropriate to use the negative control to set markers to distinguish lymphocytes, monocytes, and granulocytes.

(c) CD14 is expressed in high levels on monocytes and is dimly expressed on the mature neutrophils. Lymphocytes are negative for CD14, although a small subset may express this antigen dimly.

By combining CD45 and CD14 with the light-scatter parameters of peripheral blood cells, it is possible to identify even the minor populations of mature leucocytes. The lymphocytes express the highest levels of CD45 and little or no CD14. These cells have low forward light scatter and low right-angle light scatter. The monocytes express high levels of CD14 and intermediate levels of CD45. They are larger by forward light scatter and have increased right-angle light scatter. The neutrophils express low levels of CD14 and CD45. These cells are large by forward light scatter and have increased levels of right-angle light scatter. Eosinophils express intermediate amounts of CD45, very high right-angle light scatter, and are smaller by forward light scatter than the neutrophils. The basophils have light-scattering properties that are almost identical to those of lymphocytes. They express low levels of CD45 and low levels of CD14.

74

In setting the optimal light-scatter gate it is also necessary to have high purity, with only lymphocytes included in the gate. To identify the cells of interest (i.e. lymphocytes) it is also important to include all lymphocytes within that gate (recovery) to avoid skewing the results by omitting lymphocytes from any of the subsets. These may be conflicting goals as a large light-scatter gate is required for maximal recovery, whereas a small light-scatter gate is necessary for maximal purity.

Protocol 5

Optimizing the light-scatter gate

Method

1. Display a correlated two-parameter dot plot of CD45–FITC vs. CD14–PE fluorescence (see *Figure 7*). Set a large gate (I) around the bright CD45++, CD14– cell population to include all these cells.

2. Using this analysis gate, display a light-scatter plot of the cells within this gate (see *Figure 8*). The lymphocytes will be easily distinguished from the granulocytes that might be included in the fluorescence gate (step 1). Set a second gate (II) around the lymphocytes to determine the total number of lymphocytes in the sample. These are the cells that meet both the fluorescence and light-scatter criteria of lymphocytes.

3. Determine the light-scatter gate purity by setting gate II on light scatter and examining the CD45 vs. CD14 fluorescence (see *Figure 9*). Determine the percentage purity of lymphocytes in the light-scatter gate by comparing the number of lymphocytes to all events in that gate.

4. If the purity is less that 85%, the light-scatter gate must be reduced in size (see *Figure 10*). Determine the recovery and purity of the smaller light-scatter gate. The recovery must be >90% and preferably >95%. The purity of lymphocytes in the light-scatter gate must be >85%. Adjust the light-scatter gate to include the maximum number of lymphocytes while reducing the other contaminating cells.

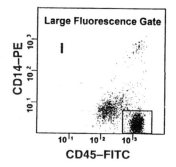

Figure 7 Display fluorescence for CD45 and CD14 and set a gate around the lymphocytes.

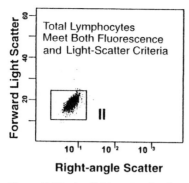

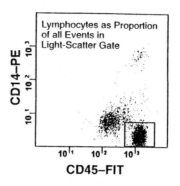

Figure 8 Display light scatter for the gated lymphocytes in *Figure 7*.

Figure 9 Determine the lymphocyte gate purity by gating on II in *Figure 8*.

**Reducing Light-Scatter
Gate to Increase
Lymphocyte Purity**

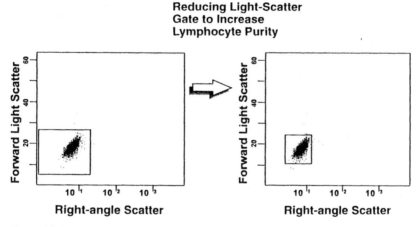

Figure 10 Deducing the size of the light-scatter gate to meet the requirement for lymphocyte purity.

9.2 Defining a reference population: T and B lymphocytes are identified using CD3/CD19

In multi-colour immunofluorescence analysis, the reference populations are not only used to discriminate positive from negative cells but also to discern single staining cells of each colour from multiple-stained cells. For this reason, a simple isotype control is not sufficient to serve as the reference population to set the fluorescence boundaries.

In the lymphocyte panel recommended by the CDC, the reference populations are established using non-overlapping T and B lymphocyte populations:

(a) CD3 is the most specific reagent for identifying T lymphocytes, whereas CD19 is the most specific reagent for identifying B lymphocytes. Using this combination of reagents, no cells should be positive for both CD3 and CD19 (see *Figure 11*). Therefore, with a CD3–FITC/CD19–PE stain, the CD3+ cells are only green, thus providing an ideal reference population against which

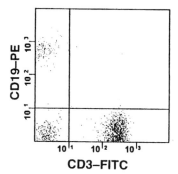

Figure 11 Plot of CD3–FITC (T cells) vs. CD19–PE (B cells) showing three discrete populations.

to compare the subsequent reagent combinations that depend on the enumeration of doubly stained cells.

(b) CD3 and CD19 define discrete separated populations. Any double-positive cells are artefacts of the sample or specimen. This is far better for setting boundaries between positive and negative cells than a simple isotype control. The single-colour populations in this tube provide the appropriate reference populations that discriminate positive cells. Also included in this combination is a population of cells that express neither CD19 nor CD3. These natural killer (NK) cells provide another double-negative reference population. (A few double-positive cells may be identified when a T cell and B cell form a doublet. In this way, two lymphocytes are stuck together and flow through the instrument as if they were one large cell.)

(c) The combination of CD3/CD19 can be used not only for setting the fluorescence boundaries, but as a quality control check for compensation and for non-specific staining. This reagent combination should display three discrete populations at right angles to each other. If the populations are not discrete or if the populations are curved (not orthogonal), a flag should be noted on the sample and/or specimen for more careful analysis.

9.3 Identifying the target cells as a discrete population: CD4+ T lymphocytes are identified using CD3/CD4 (10)

As with many monoclonal antibodies, CD4 is not completely cell-type specific as it reacts with both T-helper/inducer cells and monocytes. A second reagent is required to identify uniquely the CD4+ T lymphocytes as distinct from the monocytes:

(a) The T cells are identified using CD3 (see *Figure 12*). The double-positive cells, the CD3+/CD4+, are the true CD4-helper/inducer T lymphocytes. These are the primary targets of HIV infection. This population decreases during the disease progression and immunosuppression observed following HIV infection. Clinically, this is the most important cell population to monitor.

(b) The CD3–/CD4+ population is composed of monocytes. These cells express CD4 dimly compared with the CD4+ T lymphocytes, and are included in the

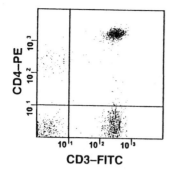

Figure 12 Plot of CD3–FITC vs. CD19–PE showing CD4+ T cells (double-stained) and CD4+ monocytes.

analysis to varying degrees based on the purity of the light-scatter gate. The combination of CD3/CD4 allows for the complete separation and identification of lymphocytes from the CD4+ monocytes. Because of this complete separation, it is better to have a large light-scatter gate that includes all the lymphocytes rather than a tight, purer lymphocyte light-scatter gate. Inclusion of some monocytes into the light-scatter gate will not affect the data since they can be easily distinguished from the CD3+/CD4+ T lymphocytes. Platelets and red cells (debris) and unstained lymphocytes are identified within the double-negative CD3+/CD4 population.

(c) This reagent combination can also be used as another quality control check for compensation and non-specific binding. The double-negative population is an internal negative control. The CD3+/CD4– population should also overlap identically with the CD3+ population in the CD3/CD19 reagent tube.

9.4 Identifying the target cells as a continuous population: CD8+ T lymphocytes are identified using CD3/CD8

The expression of CD8 is not limited to T-cytotoxic/suppressor cells since this antigen is also detected on some, but not all, NK cells. The CD8 antibody must be paired with CD3 to uniquely identify the T-cytotoxic/suppressor cells:

(a) The populations of cells that express both markers, CD3+/CD8+, are the true CD8+ T lymphocytes (see *Figure 13*) and they include both dim and bright CD8+ populations. The CD8 distribution is often continuous with the negative cells. Therefore, the proportion of CD8+ cells depends upon instrument sensitivity, establishment of correct compensations, and use of the proper reference population.

(b) The CD3–/CD8+ cells are a subset of NK cells and a few granulocytes. These cells tend to be dim CD8+ and are easily distinguished from the CD3+/CD8+ T lymphocytes by their lack of CD3.

(c) This combination of reagents can be used as an additional quality control check of compensation and non-specific staining. The double-negative population also becomes an internal negative control. The CD3+ population should appear in exactly the same spot as the two previous tubes, CD3/CD19 and CD3/CD4.

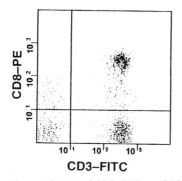

Figure 13 Plot of CD3–FITC vs. CD8–PE showing CD8+ T cells (double-stained) and CD8+ NK cells.

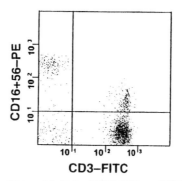

Figure 14 Plot of CD3–FITC vs. (CD16 and CD56) PE showing CD16+ and/or CD56+ NK cells and CD56+ T cells (double-stained).

9.5 Identifying all the cells of one major group: NK cells

(a) CD16 (FcRIII) is expressed on most NK cells but is also expressed on neutrophils. This antigen is lost during activation on NK cells and is expressed relatively dimly.

(b) CD56 is expressed on most, but not all, NK cells as well as some T lymphocytes. Used in combination with CD3, it is possible to distinguish between the CD3+/CD56+ T lymphocytes and the CD3–/CD56+ NK cells.

(c) The most complete identifier of all NK cells uses all three monoclonal antibodies in combination. The NK cells express either CD16 or CD56 but they do not express CD3 (see *Figure 14*). The combination of CD16 and CD56 also helps to separate the NK cells in intensity from the double-negative populations. In this way they form a discrete cluster distinct from double-negative cells.

9.6 Internal consistency of sample results for quality control: replicate CD3 determinations

The selection of the reagents for the CDC recommended panel for CD4 enumeration (see *Table 2*) provides for significant internal quality control for each specimen:

(a) Since CD3 is included in four separate tubes, replicate CD3 determinations assess any variation between tubes. The difference between the CD3 determinations should not exceed the variance defined with the laboratory. The difference between the highest and lowest values of CD3 for any specimen should not exceed 3%. If the highest and lowest values are greater than 3%, the data should be flagged for further review.

(b) If a specimen has been flagged it is usually easy to identify which of the samples is suspect. In most cases three of the four CD3 values are very close. The outlier can then be identified for more careful analysis. At this point it is important to ensure that the light scatter is identical for each sample tube. It is possible that the light-scatter pattern for the outlier tube may be different from the others. It is important to check to see that the proper reagents were added to the right sample tubes. It is also important to check the position of the fluorescent markers or boundaries between the negatives and positives. It is possible that, on the outlier sample, the boundaries are inappropriate for that particular tube.

(c) If the results are still outside the 3% specification, the outlier sample must be rejected. Providing all other quality control criteria are met, it is possible to report the results for the remaining sample tubes.

9.7 Internal consistency of sample results for quality control: accounting for all lymphocytes

The reagent panel recommended for monitoring HIV disease progression allows for the identification of all lymphoid cells—T, B, and NK cells:

(a) The sum of the %T + %B + %NK should equal 100% of the lymphocytes.

(b) T cells are defined using the average of the four CD3 counts in tubes 2–5 (*see Table 5.2*) of the panel.

(c) The B cells are identified as CD19+ and CD3– in tube number 2 of the panel.

(d) The NK cells are CD3–, but either CD16+ or CD56+ in tube number 5.

(e) Optimally, the sum of the %CD3+ plus %CD19+ plus %CD3– (CD16 plus CD56+) equals 100% ± 5% of the lymphocytes defined by CD45/CD14. Minimally, this variability should be <10%. The summation requires a correction for non-lymphocytes (purity) in the light-scatter gate.

(f) The summation of lymphocytes uses data from all five tubes. The sum will only be valid if all tubes are consistent.

10 Isotype controls

Isotype control reagents are often used to define the position of the negative cells and set the fluorescence markers (see *Figure 15*).

The use of isotype controls as a reference population assumes that the same non-specific staining and biochemical properties are present as in the test

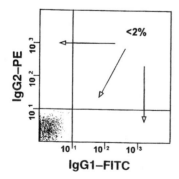

Figure 15 Two-parameter fluorescence dot plot for the isotype control tube.

reagents. It should be noted that the use of isotype controls may not exactly match each of the reagents in the panel. There may be differences in isotype (heavy chain), colour, or concentration between the control and the test reagents. The use of an isotype (double-negative) control may not be appropriate, especially for reagents that identify groups of cells that stain with two colours or more. A more appropriate control would be reagents that accurately define the green-only and the orange-only populations of cells along with a double-negative population as observed with CD3/CD19 on normal lymphocytes.

It is preferable to include the negative reference population within the panel selected, rather than to add an additional isotype control. In the recommended CDC panel, a double-negative population is observed in tubes 2–5. Another quality control step is to examine the relative positions of this negative population between all tubes. If the non-specific binding is comparable between samples, the position of this population should remain the same.

Isotope controls are less important when the reagent panel is composed entirely of reagents yielding discrete populations compared with continuous populations. Controls are required to distinguish between single-positive and double-positive cells with continuous populations.

An additional step that permits a measurement of non-specific staining is to run a tube containing unstained cells in parallel with the recommended panel. The position of the unstained cells (autofluorescence) can be compared to the populations (observed by overlaying the plots) and provides a measure of relative non-specific binding of the antibodies to the cell populations.

11 DNA content/surface marker analysis

The flow cytometric evaluation of DNA content in neoplastic processes has both diagnostic and prognostic value. Most neoplastic tissues contain both a normal component and aberrant cells that may be a minor portion of the total cell number. A DNA content analysis, in which only nuclei are tested, may yield a predominantly normal histogram with only a few 'neoplastic' stem-line events included. By combining cell-surface antigen expression with DNA content analysis, the normal cells can be excluded and the abnormal subpopulation of interest can be studied. The following protocol describes DNA content analysis combined with surface antigen quantitation.

Protocol 6

Combined measurement of DNA and a surface marker

Reagents

- FITC-conjugated antibodies
- PBS
- 50% ethanol in PBS
- 125 units/ml RNase (Sigma)
- 20 µg/ml propidium iodide

Protocol 6 continued

Method

1. Using isolated cells, centrifuge 1–2 ml of cells at 2–3 $\times$ 10^6 cells/ml. To the cell button add the appropriate amounts of monoclonal antibodies for surface marking.

2. Incubate this suspension for 20 min at room temperature in the dark. Wash the cells twice using cold PBS and pellet them at 300 g.

3. To the cell pellet, add dropwise with vortexing 1 ml of cold ($-20\,°C$) 50% ethanol in PBS. Fix for 30 min or leave overnight.

4. After fixation, spin down the cells and aspirate the ethanol/PBS leaving about 50 µl of fluid above the cells. Add 500 µl PBS to the pellet along with 200 µl RNase. Incubate on ice for 30 min.

5. Wash the cells twice with PBS. Add 1 ml propidium iodide to the pellet and incubate for at least 1 h in the dark at 4°C. At this point the cells are ready for analysis.

References

1. Loken, M. R. and Stall, A. M. (1982). *J. Immunol. Methods*, **50**, 85.
2. Parks, D. and Herzenberg, L. (1984). *Methods in Enzymology*, **108** , 197.
3. Terstappen, L. W. M. M., Mickaels, R., Dost, R., and Loken, M. R. (1990). *Cytometry*, **11**, 506.
4. Stewart, C. S. and Stewart, S. J. (1999). *Cytometry*, **38**, 161.
5. Loken, M. R. (1997). In *Current protocols in cytometry* (ed. P. A. Robinson), Unit 10.6. Wiley, New York.
6. Loken, M. R., Brosnan, J., Bach, B., and Ault, K. (1990). *Cytometry*, **11**, 453.
7. Jackson, A. (1990). *Clin. Immunol. Newsletter*, **10**, 43.
8. Carter, P. H., Resto-Ruiz, S., Washington, G. C., Ethridge, S., Palini, A., Vogt, R., *et al.* (1992). *Cytometry*, **13**, 68.
9. National Committee for Clinical Laboratory Standards (1992). *Clinical applications of flow cytometry: quality assurance and immunophenotyping of peripheral blood lymphocytes*. National Committee for Clinical Laboratory Standards (H42-T), Villanova, PA.
10. Centers for Disease Control (1992). *Morbid. Mortal. Week. Rep.*, **41**, 1.
11. Giorgi, J. and Hultin, L. (1990). *Clin. Immunol. Newslett.*, **10**, 55.

Chapter 6
Analysis of DNA—general methods

M. G. Ormerod

34, Wray Park Rd, Reigate, Surrey RH2 0DE, UK

1 Introduction

The measurement of DNA was one of the first and is still one of the most widespread applications of flow cytometry. Malignant cells are frequently aneuploid and there have been many studies of the possible prognostic significance of ploidy in human tumours. The content of DNA also gives information about the cell cycle, so that this measurement can be of value in cell biology. In particular, it can yield useful information about the action of cytotoxic drugs. The measurement of DNA may also be combined with estimation of an antigen, so that the expression of a particular protein through the cell cycle may be measured or, in a cell mixture, the ploidy or cell cycle of a particular class of cell estimated (see Chapter 5, Section 11, and Chapter 9).

All the established methods use dyes which bind specifically and stoichiometrically to nucleic acids and whose fluorescence is enhanced on binding. The properties of the more important of these compounds are described in Chapter 2. A variety of methods have been published for measuring the DNA content of cells. Those which have been adapted for use in my laboratory are given in this chapter. However, measurement of the DNA content of a cell on its own only gives a static view of the cell cycle. A more dynamic picture can be obtained either by pulse-labelling cells with 5-bromodeoxyuridine (BrdUrd) and using an antibody to BrdUrd (see Chapter 10) or by continuously labelling with BrdUrd and staining nuclei with Hoechst 33258 (see Chapter 11). Because of the importance of the measurement of DNA in looking at the cell cycle, its main features are described briefly in the next section.

While this chapter concentrates exclusively on mammalian cells, the methods described can be, and have been, adapted for other organisms (1, 2).

2 Definitions

2.1 The cell cycle

A quiescent cell, which is not growing or involved in cell division, is often referred to as being in the G_0 state. If division is triggered, the cell first enters

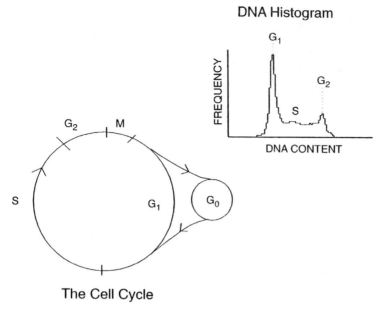

DNA Histogram

Figure 1 A diagram illustrating the cell cycle together with a DNA histogram.

the G_1 phase of the cell cycle, during which the amount of RNA increases and certain proteins essential for DNA replication are made. When the cell starts to make new DNA it has entered the S, or DNA synthetic, phase of the cycle. During this phase, the DNA content of the cell increases until it has doubled. At this point, DNA synthesis ceases and the cell is in the G_2 phase. Finally the cell will enter mitosis (M) and divide returning to G_1, if cell division is to be sustained, or to G_0. Cells in G_2 and M phases of the cell cycle have double the DNA content of those in G_0 and G_1. Cells in S phase will have a DNA content lying between these extremes. If the DNA content of, say, 20 000 cells is measured and a histogram of the number of cells against DNA content is plotted, this will reflect the state of the cell cycle (see *Figure 1*).

2.2 Ploidy

The number of chromosomes in a germ cell (gamete) is called the haploid number, n, for that species and the number in a somatic cell, the diploid number, $2n$. Occasionally some somatic cells may be tetraploid ($4n$) or even octaploid ($8n$). The number of chromosomes in a tumour is frequently greater than $2n$ (hyperdiploid) and sometimes less (hypodiploid). An abnormal number of chromosomes is called aneuploidy and this is reflected by a change in the content of DNA.

When aneuploidy is measured as a change in DNA content, as opposed to a change in the chromosomes, it should be referred to as DNA aneuploidy. The DNA content of a tumour may be expressed as the DNA index (DI), defined as the ratio between the DNA content of a tumour cell and that of a normal diploid cell.

3 General comments

3.1 Preparative methods

The object of all the preparative methods is to obtain either intact nuclei or single cells with the minimum degradation of their DNA. If cells are used, generally they must be fixed or permeabilized to allow access of the dye. Furthermore, many dyes bind to both DNA and double-stranded RNA so that the latter must be removed by digestion with ribonuclease. Cleanliness of all glassware and sterility of the solutions used are important. Traces of contaminating deoxyribonucleases can degrade the nuclear DNA and hence affect the quality of the analysis.

3.2 Analysis

It is important during the analysis to ensure that the profiles are as sharp as possible and that only single cells or nuclei are measured (that is, two or more cells stuck together are excluded).

3.2.1 Quality of the histogram

The quality of a DNA histogram is estimated from the width of the peak of DNA from cells in G_1 of the cycle. This is measured by the coefficient of variation (CV) across the peak and is calculated from the standard deviation (SD):

$$CV = 100 \times SD/(peak\ channel)\%.$$

The smaller the CV of the peaks in the histogram of DNA content, the more accurate is the measurement of ploidy and the better the estimation of the percentage of cells in the different compartments of the cell cycle. It is essential that any unnecessary broadening of the peaks due to misalignment of the instrument should be eliminated.

The performance of the instrument should be checked daily using fluorescent beads of known CV; these can be purchased from several manufacturers. The CV and the peak channel number for a standard set of conditions (laser power, PMT voltage and gain) should be recorded. If these fall outside predetermined limits (for example, 2% CV), action should be taken to restore the instrument's performance.

- Check that the flow rate has not been accidentally set too high.
- Check that there is not a partial blockage of the flow cell.
- If possible (with a conventional cell sorter), realign the instrument.
- If realignment is not possible (most bench-top instruments), call in the service engineer.

Any perturbation of the sample stream in the cytometer will increase the CV, and for this reason the concentration of cells or nuclei should be kept high (between 5×10^5 and 2×10^6/ml) and the flow rate low.

3.2.2 Eliminating clumps of cells

If two nuclei or cells in G_1 of the cell cycle are stuck together they will have the same DNA content as a single cell in G_2, and the two must be distinguished if the DNA histogram is to reflect accurately the state of the cell cycle. Although the number of clumps can be reduced by careful sample preparation and by passing the sample through a syringe needle, it is an advantage to use an instrument that can focus the laser beam down to an ellipse with a cross-section of about 10 μm or less. The difference can then be resolved by analysing the shape of the signal generated as the particle crosses the laser beam (3). The width of the signal generated by a fluorescent particle in the flow cytometer is given by the sum of the width of the laser beam and the particle diameter. As a cell progress through its cycle, while the content of DNA doubles, the nuclear diameter does not increase by more than approximately 25%. Consequently, there is only a small increase in the signal width. Because of the nature of the flow system, two particles stuck together will align one behind the other; their combined diameter will be at least two nuclear diameters. This is illustrated in *Figure 2*.

A practical demonstration is given in *Figure 3*, which shows data recorded using a Beckman–Coulter Elite. On this instrument, the width of the signal is best measured by comparing the signal height (peak) to its area. These parameters can be displayed on a cytogram in which the clumps can be distinguished from single cells and a region selected to include only the latter.

It can be seen in *Figure 3* that the single cells fall on a diagonal line. An alternative is to display the ratio between the peak and area of the DNA fluorescence signal. Clumps of cells will have a lower ratio. This ratio is sensitive to the shape of the nuclei and may reveal nuclear changes, for example compaction of the nuclear chromatin (see Chapter 14, *Figure 3*).

Figure 4 shows a similar analysis performed on a Becton Dickinson FACScan.

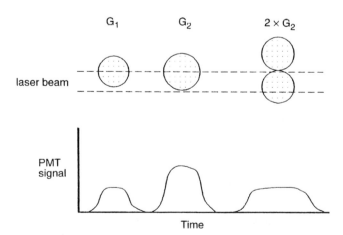

Figure 2 The signals generated when a cell in G_1, a cell in G_2, or two cells in G_1 of the cell cycle cross the laser beam.

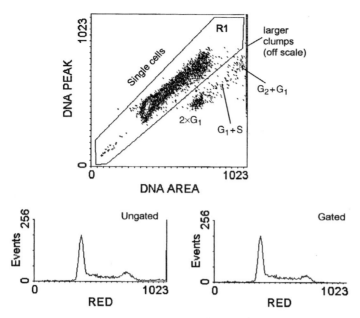

Figure 3 The exclusion of clumped cells from an analysis of DNA (method 1). A cytogram of the peak vs. the area of the DNA signal is displayed. The single cells are enclosed in a region (R1). Clusters consisting of clumps of cells are marked. The effect of gating on the cytogram is shown in the two DNA histograms. Propidium-iodide stain of ethanol-fixed cells (human lymphoblastoid cell line, W1L2). Beckman–Coulter Elite ESP.

A cytogram of signal width vs. area is displayed. Again there is a clear separation of single and clumped cells.

On instruments in which the laser beam is focused to a circular spot of diameter 60 μm or more, this type of analysis is not possible. However, the presence of clumps can be detected independently of the configuration of the instrument by inspecting the DNA histogram in the region corresponding to a DNA content of three times that of the cells in G_1/G_0. A peak of cells here arises from three cells in G_1 or a cell in G_2 stuck to a cell in G_1. Some analysis software uses this peak to estimate the number of G_1 doublets under the G_2 peak and to subtract them from the analysis.

3.2.3 Standardization

When measuring ploidy, the concentration of cells from one sample to another should be kept approximately the same so that the dye:DNA ratio does not vary too much. If a particular sample has a much higher number of cells than the others being measured, dilute it with dye solution and re-run the sample. A standard sample is also needed. The absolute amount of propidium iodide (PI) bound to DNA is affected by the conditions of sample preparation, particularly fixation. It is therefore important that the standard undergoes the same process-ing as the sample under study. When samples of cancerous tissue are studied, there are usually some normal diploid cells present and these can act as an

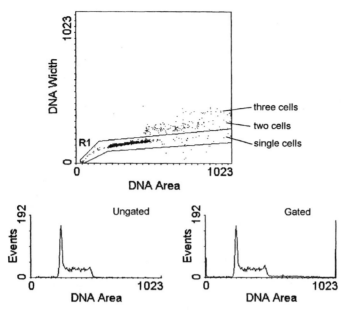

Figure 4 The exclusion of clumped cells from an analysis of DNA (method 2). A cytogram of the width vs. the area of the DNA signal is displayed. The single cells are enclosed in a region (R1). Clusters consisting of clumps of cells are marked. The effect of gating on the cytogram is shown in the two DNA histograms. Propidium-iodide stain of ethanol-fixed cells (murine cell line, L1210). Becton-Dickinson FACSort. Data recorded during the Royal Microscopical Society's Course on Flow Cytometry, Cambridge, 1994.

internal standard. If a DNA histogram is recorded from nuclei extracted from a paraffin block (see Section 4.3 and *Figure 7* below), such an internal standard is all that can be used. When using fresh tissue or cultured cells, either chicken or trout erythrocytes, or preferably both, may be used (4) (see *Figure 5*).

On most instruments, it will be found that the ratio of the channel numbers of the DNA peaks from G_1 and G_2 cells is not exactly two. This reflects the baseline adjustment of the amplifier and, if the ratio is unacceptable (say, outside the range of 1.9 to 2.1), the amplifier should be readjusted by the service engineer.

Consensus documents on the standardization of DNA flow cytometry in clinical pathology have been published (5, 6).

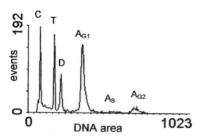

Figure 5 DNA histogram from a fine-needle aspirate of human breast carcinoma prepared according to the Vindelov method with added chicken (C) and trout (T) erythrocytes. The position of the diploid (D) and the aneuploid (A_{G1}, A_S, and A_{G2}) cells are marked. Data recorded on a Becton-Dickinson FACScan and supplied by Ib Jarle Christensen, The Finsen Laboratory, Copenhagen.

4 Experimental methods

4.1 Cultured cells and cells in suspension

4.1.1 Fixing the cells

A suitable method for staining with propidium iodide is given in *Protocol 1*. The cells are fixed in ethanol and the cells can be stored in this form at 4°C for at least a month. If required, several samples from an experiment may be collected and run together. The disadvantage is that the cytoplasm is left intact and, if an accurate estimation of ploidy is required, particular care must be taken to ensure that all the double-stranded RNA has been removed by enzymatic digestion.

Protocol 1

Measurement of DNA of cells in suspension using ethanol fixation

Equipment and reagents

- PI stock solution: 400 μg/ml PI (Sigma or Molecular Probes) in water. Store at 4°C.
- RNase solution. Prepare just before use by dissolving 1 mg/ml RNase (Sigma) in PBS.

Method

1. Prepare a suspension of single cells (see Chapter 3) in 200 μl of PBS.

2. Add vigorously 2 ml of ice-cold 70% ethanol, 30% distilled water. Leave for at least 30 min on ice.

3. Harvest the cells by centrifugation (300 g for 5 min). Resuspend in 400 μl of PBS, pH 7.3.

4. Check the cells microscopically. If they are badly clumped, pass through a 25-gauge syringe needle.

5. Add 50 μl of the RNase and 50 μl of the PI solutions. Incubate at 37°C for 30 min.

6. Analyse using an argon-ion laser tuned to 488 nm and measuring forward and orthogonal light scatter and red fluorescence (where possible measuring area and either peak or width of the fluorescent signal).

Occasionally it is found that the cells are badly clumped after fixing in ethanol. This can be caused by the presence of DNA from dying cells in the suspension before fixation. If this is a problem, before fixation, suspend the cells in culture medium (for example, RPMI from Gibco) containing 4 μg/ml DNase (Type I bovine pancreatic DNase, Sigma) and incubate at 37°C for 10 min. Moderate clumping can often be improved by passing the fixed cells through a 25-gauge hypodermic needle.

4.1.2 Lysing the cells

For cells with a large cytoplasmic to nuclear ratio, it may be preferable to lyse the cells rather than to fix them. Make up a stain–detergent solution by dissolving

1 g trisodium citrate, 564 mg NaCl, 300 μl Nonidet P-40 (Sigma), and 10 mg propidium iodide in 1 litre of distilled water. Just before use, dissolve 1 mg of RNase in 100 ml solution. Prepare a suspension of single cells, centrifuge them, discard the supernatant, and resuspend the pellet in ice-cold, stain–detergent solution. Analyse using an argon-ion laser tuned to 488 nm and measuring red fluorescence. The cells should not be stored for more than a few hours in this form.

4.1.3 Viable cells

If it is necessary to maintain cell viability, the cells can be stained with the *bis*-benzimidazole, Hoechst 33342 dye (see Chapter 2). The dye is specific for DNA and is taken up actively by cells, the rate of uptake varying from one type of cell to another. Hoechst 33342 is excited by UV light and gives a blue fluorescence. As the amount of dye bound to the DNA increases, the fluorescence decreases and shifts towards the red. When cells are incubated with dye, the amount of fluorescence at first increases and then decreases with the time of incubation. Therefore incubation conditions need to be carefully determined for each type of cell. Incubation should be at 37°C and, typically, the dye might be used at 10 μg/ml with an incubation time of between 5 and 40 min.

Before analysis add PI at 5 μg/ml to stain dead cells. Tune the laser to give UV and record light scatter and red (area), blue (area plus either peak or width) fluorescences. Use pulse shape analysis of the blue fluorescence to eliminate clumps, red (PI) fluorescence to gate out dead cells, and display a histogram of blue fluorescence (*Figure 6*).

4.2 Measurement of the DNA content of fresh tissue

There are several methods for extracting nuclei from fresh tissue; Chapter 3 gives methods for dispersing solid tissues. I have used a method adapted from that of Petersen (7) (see *Protocol 2*).

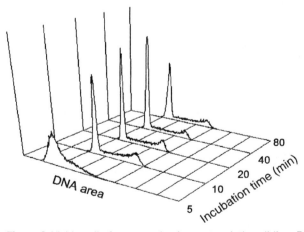

Figure 6 Viable cells from a murine haematopoietic cell line, BAF3, stained with 10 μg/ml Hoechst 33342 for the times in minutes indicated on the figure at 37°C. PI (5 μg/ml) was added before analysis and dead cells excluded by gating on red fluorescence. Argon-ion laser outputting UV (360 nm).

Protocol 2

Measurement of DNA in nuclei from fresh tissue

Equipment and reagents

- Stain–detergent solution: 1 g trisodium citrate, 564 mg NaCl, 300 μl Nonidet P-40 (Sigma), 10 mg propidium iodide in 1 litre of distilled water. Just before use, dissolve 1 mg of RNase in 100 ml solution.
- Stainless-steel, tea-strainer
- 60 μm pore, nylon mesh (Lockertex or Small Parts)

Method

1. Take 200 mg wet weight or more of tissue. Mince with scalpels in tissue culture medium.

2. Filter through a stainless-steel, tea-strainer.

3. Centrifuge at 800 g for 5 min.

4. Aspirate and discard the supernatant. Resuspend the pellet in 10 ml of stain-detergent solution.

5. Mix on a rotator at 4 °C for 1–4 h. The exact time will depend on the tissue used and must be found by trial and error.

6. Filter through 60 μm pore, nylon mesh.

7. Analyse using an argon-ion laser tuned to 488 nm and measuring forward and orthogonal light scatter and red fluorescence (where possible measuring area and either peak or width of the fluorescent signal).

Note: If this method gives a low yield, additional treatment with trypsin (Sigma) may be necessary to achieve better dispersion of the tissue (see also ref. 4). The exact conditions need to be determined for the particular tissue under study. Typically, one might use 1 mg/ml for 10 min at 37 °C.

4.3 Measurement of the DNA content of fixed tissue embedded in paraffin wax

Tissue removed during an operation or an autopsy is often fixed and embedded in paraffin wax prior to cutting tissue sections for staining and histopathological examination. These blocks are stored for many years and the development of a method for extracting nuclei from them has given flow cytometrists access to much archival material (8). The method described in *Protocol 3* uses xylene to extract the paraffin from the sections. Step 2 should be performed in a fume cupboard. Some workers prefer to use Histoclear (National Diagnostics) as an alternative to xylene, thereby avoiding its possible toxicity.

On analysis, there will be a certain amount of degraded material present. This is inevitable, as some nuclei will have been sliced when the sections were cut and have less DNA in them. This makes an attempt to measure the cell-cycle parameters less accurate. Occasionally, the whole profile is heavily degraded. This defect can seldom be remedied as it usually reflects the way the tissue has been handled before and during embedding in paraffin. Sometimes light scatter can be used to exclude the more badly degraded material. Light scatter can also often be used to distinguish normal from malignant cells, the latter often have higher right-angle light scatter, particularly in the case of breast carcinomas (9). *Figure 7* shows an example in which a diploid subpopulation has been resolved using light scatter.

Protocol 3

Measurement of DNA in fixed tissue embedded in paraffin wax

Equipment and reagents

- Small biopsy (curetting) cassettes (Raymond Lamb)
- 23-gauge needle
- Pepsin solution (freshly made): 0.9% NaCl, 0.5% pepsin adjusted to pH 1.5 with HCl
- PBS/PI/RNase: 100 μg/ml RNase, 50 μg/ml PI in PBS. Store a PBS/PI solution at 4°C. Add the RNase just before use.
- 0.1 M phosphate buffer, pH 7.0

Method

1. Cut two or three 50 μm sections from each block. Place the sections in a small biopsy cassette. Store in xylene in glass containers.
2. Agitate well and, in a fume cupboard, rinse in xylene to ensure complete removal of the paraffin wax.
3. Wash the tissue twice in ethanol, once in 50% (v/v) ethanol, and twice in PBS.
4. If the tissue is particularly fibrous, incubate for 1 h at 37°C in 1 mg/ml collagenase in PBS.
5. Remove the tissue carefully from the cassette and place in a tube suitable for centrifugation. Add the pepsin solution. Incubate at 37°C for 1 h.
6. Pass through a 23-gauge needle to break up clumps. Wash by centrifugation (300 g for 5 min) once in 0.1 M phosphate buffer, pH 7.0 and once in PBS.
7. Finally resuspend in PBS/PI/RNase and incubate for 30 min at 37°C.
8. Analyse using an argon-ion laser tuned to 488 nm and measuring forward and orthogonal light scatter and red fluorescence (where possible measuring area and either peak or width of the fluorescent signal).

The amount of PI bound may be affected by the conditions of fixation of the original tissue (10). For this reason, the best standard for measuring a DNA index is given by normal diploid cells within the sample. There are usually some normal cells present in a tumour biopsy. As routine, sections for conventional histological staining (haematoxylin and eosin) should be taken immediately pre-

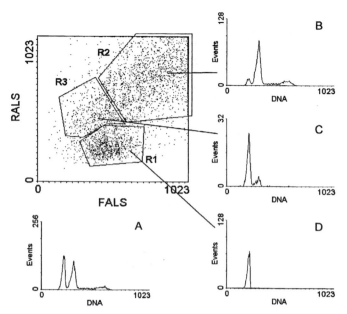

Figure 7 Nuclei extracted from a formalin-fixed, paraffin-embedded human breast tumour and stained with PI. The cytogram shows the light scatter. Regions were set as shown, and the DNA histograms for all the nuclei and for nuclei falling within the different regions are shown. Nuclei with low right-angle light scatter (RALS) (region R1) were from normal cells; those with high RALS (regions R2 and R3) from tumour cells. The channel numbers of the major peaks are shown. (A) Ungated data; (B) gated on high RALS and high FALS (R2) (aneuploid tumour); (C) gated on high RALS and low FALS (R3) (diploid tumour); and (D) gated on low RALS (R1) (diploid normal). Data recorded on a Becton-Dickinson FACSort during a Royal Microscopical Society course in Cambridge, 1994. Material kindly supplied by Dr Richard Camplejohn, Richard Dimbleby Department of Cancer Research, London.

ceding and following the thick sections cut for flow cytometry. These sections should be examined to check for the presence of tumour in that part of the paraffin block.

Hedley (11) has published a comprehensive review of this method and the results obtained on different tumours.

4.4 Combined analysis of an antigen and DNA

Suitable protocols are given in Chapter 5, *Protocol 6* (surface antigen) and Chapter 9 (cytoplasmic and nuclear antigens).

5 Deconvolution of the DNA histogram

5.1 Introductory remarks

The most accurate method of measuring the relative numbers of cells in different stages of the cell cycle is to pulse-label with BrdUrd and to use an antibody to BrdUrd (see Chapter 10). However, this is a lengthy procedure and labelling

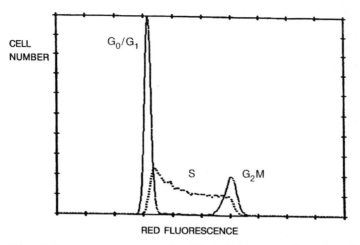

Figure 8 The separated components of the DNA histogram. The cells in S phase have been separated from those in the rest of the cell cycle using a BrdUrd label (see Chapter 10 and ref. 15) and the two parts then superimposed back on one another to demonstrate the overlap between S and the G phases.

the living cells may not always be feasible. Information about the cell cycle must often be derived directly from the DNA histogram.

The problem of deconvoluting the DNA histogram into its component parts is that of separating the DNA content of cells in early S from those in G_1/G_0 and those in late S from those in the G_2/M phase of the cell cycle. The extent of this overlap is demonstrated in *Figure 8* in which the S phase has been separated

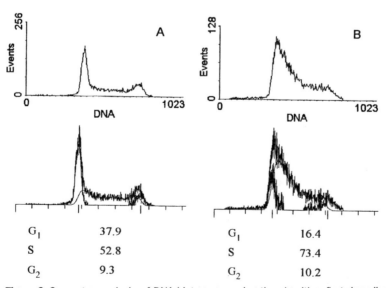

Figure 9 Computer analysis of DNA histograms using the algorithm first described by Watson *et al.* (13) and modified by Ormerod *et al.* (15) and Terry Hoy, University of Wales College of Medicine, Cardiff. Human lymphoblastoid cell line, W1L2. (A) Untreated cells; (B) cells incubated for 24 h with a thymidylate synthase inhibitor.

from G_1 and G_2 using a BrdUrd label (see Chapter 10). This figure also demon-strates the shape of the DNA histogram from normally cycling S-phase cells.

It is usually assumed that the cells in the G_1 and G_2 phases are distributed normally so that the distributions can be described by Gaussian curves. In practice, this proves to be a reasonable assumption. If the distribution of the DNA content of the cells in S phase could also be described by a simple mathematical formula, then the procedure of deconvoluting the DNA histogram would be straightforward. Unfortunately this is not the case, and the algorithms for the deconvolution use different approaches to handling the S phase or the S/G interfaces (12–15). A typical analysis of two DNA histograms is shown in *Figure 9*.

5.2 A simple method

If the S phase is not too high (>15%) and evenly distributed, a simple way of obtaining an estimate of the cell-cycle phases is to measure the number of S-phase cells in the centre of the histogram and to assume that this represents half of S phase.

Protocol 4

A simple estimate of S phase

Method

1. Calculate a lower channel from the formula: G_1 peak + (G_2 peak – G_1 peak)/4.

2. Calculate an upper channel from the formula: G_2 peak – (G_2 peak – G_1 peak)/4.

3. Measure the number of cells contained between these channels (i.e. in the centre of the histogram) and double it. This number equals the number of cells in S phase, N_S.

4. Measure the number of cells in channels less than the lower channel calculated in step 1. Subtract $N_S/4$. This value equals the number of cells in G_1.

5. Measure the number of cells in channels greater than the upper channel calculated in step 2. Subtract $N_S/4$. This value equals the number of cells in G_2.

This method is equivalent to fitting a rectangle to the S phase. This simple method does not require a sophisticated computer program—just the ability to set regions on a histogram and record the number of cells falling within the region. The analysis of a set of DNA histograms from fine-needle aspirates of human breast carcinoma by this method was compared to the analysis using a sophisticated computer program; there was no significant difference between the two sets of results (J. C. Titley and M. G. Ormerod, unpublished data). This method should not be used with cultured cells in which the S phase is often >20% and also may be perturbed (for example, see *Figure 9*).

5.3 More sophisticated methods

There are a variety of computer programs available for the analysis of DNA histograms. Often they offer a choice of algorithms. Each algorithm makes different assumptions and may give slightly different results. When analysing a set of histograms, it is important to use the same program and the same algorithm for the whole data set.

Some programs have additional features. Most will calculate the number of cells in a 'sub-G_1' peak—a measure of apoptosis, see Chapter 14. They may attempt to calculate the number of aggregated cells present. They may also attempt to subtract debris from the DNA histogram and, in material from paraffin blocks, correct for the presence of sliced nuclei. More controversially, some will analyse overlapping cell cycles. In my view, it is always preferable to concentrate on cell preparation and the correct analytical procedures rather than rely on a computer program. For example, aggregates should be eliminated during analysis using pulse-shape analysis (see Section 3.2.2). If one of the more advanced features of a computer program is used, it is important that the correct gating strategy is used. If debris subtraction is used in the computer analysis, then care should be taken to ensure that none of the debris is excluded from the DNA histogram when gating on a peak/area plot; light-scatter gating should not be used. Estimates of S phases in polyploid samples containing overlapping DNA histograms may be unreliable and should be treated with caution.

While it is useful to have an estimate of the number of cells in the different phases of the cell cycle, the most valuable information can usually be gained by a visual inspection of the histograms. For example, in *Figure 9*, the computer program correctly reflects an increase in the percentage of cells in S phase. However, visual inspection reveals the most important feature—there is an accumulation of cells in early S phase.

Acknowledgement

I thank Dr Ib Jarle Christensen, The Finsen Laboratory, Copenhagen, Denmark, for supplying the data file from which *Figure 5* was constructed.

References

1. Dien, B. S., Peterson, M. S., and Srienc, F. (1994). In *Methods in Cell Biology*, Vol. 42 (ed. Z. Darznynkiewicz, J. P. Robinson, and H. A. Crissman), p. 457. Academic Press, San Diego, CA.
2. Steen, H. B., Jernaes, M. W., Skarstad, K., and Boye, E. (1994). In *Methods in Cell Biology*, Vol. 42 (ed. Z. Darznynkiewicz, J. P. Robinson and H. A. Crissman), p. 477. Academic Press, San Diego, CA.
3. Sharpless, T. K. and Melamed, M. (1976). *J. Histochem. Cytochem.*, **4**, 257.
4. Vindelov, L. L., Christensen, I. J., and Nissen N. I. (1983). *Cytometry*, **3**, 328.
5. Shankey, T. V., Rabinovitch, P. S., Bagwell, B., Bauer, K. D., Duque, R. E., Hedley, D. W., Mayall, B. H., and Wheeless, L. L. (1993). *Cytometry*, **14**, 472.

6. Ormerod, M. G., Tribukait, B., and Giaretti. W. (1998). *Anal. Cell. Path.*, **17**, 103.

7. Petersen, S. E. (1985). *Cytometry*, **6**, 452.

8. Hedley, D. W., Friedlander, M. L., Taylor, I. W., Rugg, C. A., and Musgrove, C. A. (1983). *J. Histochem. Cytochem.*, **31**, 1333.

9. Ormerod, M. G., Titley J. C., and Imrie, P. R. (1995). *Cytometry*, **21**, 294.

10. Schutte, B., Reynders, M. M. J., Bosman, F. T., and Blijham, G. H. (1985). *Cytometry*, **6**, 26.

11. Hedley, D. W. (1989). *Cytometry*, **10**, 229.

12. Dean. P. N. (1990). In *Flow cytometry and cell sorting* (ed. M. R. Melamed, T. Lindmo. and M. L. Mendelsohn), p. 415. Wiley–Liss, New York.

13. Watson, J. V., Chambers, S. H., and Smith, P. J. (1987). *Cytometry*, **8**, 1.

14. Rabinovitch. P. R. (1994). In *Methods in Cell Biology*, Vol. 41 (ed. Z. Darnzynkiewicz, J. P. Robinson and H. A. Crissman), p. 263. Academic Press, San Diego, CA.

15. Ormerod, M. G., Payne, A. W. R., and Watson, J. V. (1987). *Cytometry*, **8**, 837.

Chapter 7
Further clinical applications

Terry Hoy,[*] Steve Garner,[†] Brian K. Shenton,[‡]
Alison E. Bell,[‡] Mark W. Lowdell,[**] John Farrant,[††]
Margaret North,[††] and Carrock Sewell[§§]

[*]University of Wales College of Medicine, Department of Haematology, Heath
Park, Cardiff CF4 4XN, Wales

[†]National Blood Service East Anglia Centre; and Division of Transfusion
Medicine,University of Cambridge, Long Road, Cambridge CB2 2PT, UK

[‡]Department of Surgery New Medical School, Newcastle-upon-Tyne NE2 4HH, UK

[**]BMT Unit, Deptartment of Academic Haematology, The Royal Free Hospital
Medical School, Pond Street, London NW3 1YD, UK

[††]Antigen Presentation Research Group, Northwick Park, Institute of Medical
Research, Watford Road, Harlow HA1 3UJ, UK

[§§]Department of Immunology, Royal Free Hospital School of Medicine, Rowland
Hill Street, London NW3 2PF, UK

1 Introduction

Flow cytometers are now to be found in many clinical laboratories where their
primary use is for immunophenotyping. This chapter contains a selection of
other applications that have proved useful in the clinical environments of
haematology, blood transfusion, transplantation, and immunology.

2 The enumeration of reticulocytes in peripheral blood

The introduction of Thiazole Orange in 1986 (1) solved many of the problems
associated with measuring reticulocytes by flow cytometry (2) and it rapidly became the fluorochrome of choice. Its fluorescence increases by a factor of 3000
on binding to RNA, which implies a high signal-to-noise ratio giving a reliable
reticulocyte count and maturity index (3). The method can be applied to any
flow cytometer with a suitable laser (488 nm) and analytical filters (e.g. a standard FITC bandpass), the absorption maximum being at 509 nm and emission
maximum at 533 nm.

Protocol 1

Enumeration of reticulocytes in peripheral blood

Reagents

- Diluent: PBS with 2 mM EDTA and 0.02% sodium azide
- Working solution: dilute the stock solution 1 in 10 000 with diluent. Can be stored for 1 week if kept in the dark at room temperature.
- 1 mg/ml stock solution of Thiazole Orange (Molecular Probes) in methanol. Stable at 5 °C for long periods.
- EDTA- or heparin-anticoagulated whole blood

A. Preparing samples

1. Add 5 µl of blood to 1 ml of the working solution and a further 5 µl of blood to 1 ml of the diluent as a control.

2. Incubate at a fixed temperature (20 °C) for 1 h.

B. Acquiring data on the cytometer

1. Analyse on a flow cytometer within 100 min of starting the incubation.

2. Use logarithmic amplification on forward and side scatter to help delineate red cells from white cells and platelets.

3. Collect sufficient data for the precision required. Normal reticulocyte counts can be as low as 0.5% of the red blood cells, 50 000 events will be required to produce coefficients of variation (CVs) below 10%.

C. Analysing the data (see Figure 1)

1. Display forward vs. side scatter and place a region around the red cells, these have a characteristic horseshoe shape after logarithmic amplification (see *Figure 1A*).

2. Using this region as a gate, overlay histograms of the test and control data (see *Figure 1B*).

3. Note that the test histogram for the majority of events may be slightly displaced, this is caused by background staining of the mature red cells which may be enhanced if overstained.

4. Examine the high fluorescence region and eliminate any very bright events, these can arise from nucleated red cells in the sample (see *Figure 1C*).

5. Determine the percentage reticulocytes between appropriate markers placed above the mature red cells and below any nucleated cells.

6. If required, derive a maturity index (ref. 3 and see *Figure 1C*).

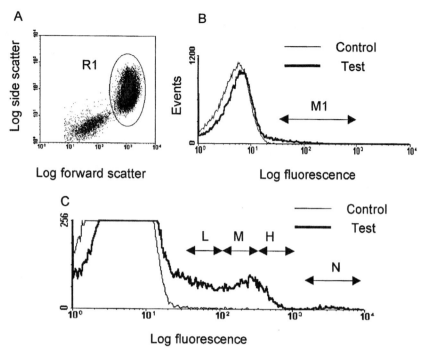

Figure 1 Analysis of reticulocytes. (A) Region R1 defines the red cells and reticulocytes, the smaller population being platelets. (B) Histograms of Thiazole Orange fluorescence gated on region R1 for a control and test sample. The marker, M1, defines 0.5% for the control and 2.5% for the test sample, giving the reticulocyte percentage as 2% of the red cell population. (C) Histograms from a sample with a high reticulocyte count allocated to regions of high (H, 2.3%), medium (M, 4.7%), and low (L, 4.9%) fluorescence corresponding to increasing maturity. Some nucleated red cells are present, defined by marker N (0.5%).

3 The enumeration of CD34+ cells in peripheral blood stem-cell harvests and bone-marrow harvests

Peripheral blood stem-cell transplants are now a treatment option for a number of malignancies (4). The patient's haematopoietic system is stimulated to produce pluripotent cell precursors (stem cells) which exit the bone marrow and circulate in the peripheral blood. These cells can be harvested and frozen and the procedure may be repeated a number of times. The patient is subsequently treated with chemotherapy to destroy the diseased bone marrow, after which the harvested stem cells are re-infused to repopulate the bone marrow allowing haematopoietic recovery to take place.

The number of stem cells re-infused must be above a certain threshold to enable a successful marrow recovery. Accurate enumeration of the percentage of stem cells in each harvest is therefore important and can be measured by flow cytometry using a combination of CD34 and CD45 monoclonal antibodies (5).

CD34+ cells in harvested bone marrow can also be estimated if these cells are to be used for bone-marrow transplantation.

101

Protocol 2

Enumeration of CD34+ cells

Reagents

- Peripheral blood or a bone-marrow harvest (usually taken and supplied in acid citrate dextrose)
- Isotype–PE control (optional, see below)
- Red-cell lysing solution (e.g. FACS Lysing solution, Becton Dickinson, or an ammonium chloride method; see Chapter 3, *Protocol 3*)
- CD34–PE (avoid all class I antibodies and class II–FITC conjugates)
- CD45–FITC (Use antibodies that detect all isoforms and glycoforms)
- Phosphate buffered saline with 0.5% BSA (PBSA)
- 2% paraformaldehyde solution

A. Preparing samples

1. Dilute an aliquot of the sample with PBSA to give a cell count of about 2×10^{10}/litre. Prepare at least 100 μl of diluted sample for testing.

2. Label the tubes as follows: CD45/CD34 and CD45/IgG1 (optional control).

3. Using appropriate amounts of antibody as recommended or determined by titration, add anti-CD45–FITC to both tubes, anti-CD34–PE to one tube, and isotype control–PE to the optional tube.

4. Add 50 μl of the diluted cell suspension to each of the tubes. Mix gently and incubate at 4°C for 20 min, mix again after 10 min.

5. Add 1 ml of diluted lysing solution to each of the tubes to lyse red cells. Mix the tubes well or vortex. (Alternatively, use an ammonium chloride procedure).

6. Incubate the tubes for a further 5 min at room temperature.

7. Centrifuge at 300 g for 5 min.

8. Remove the supernatant, flick the tubes to resuspend the cells, add 3 ml of PBSA and centrifuge at 300 g for 5 min. Repeat this stage

9. Resuspend the cells and add 300 μl of the 2% paraformaldehyde solution.

10. Leave the at 4°C for 10 min before analysing on a cytometer for best results.

B. Acquiring data on the cytometer[a]

1. Perform necessary checks for cytometer performance and compensation (see Chapter 5).

2. Dilute the stained samples with PBS to approximately 0.5–1 ml.

3. Collect 50 000 events for the CD45/CD34 tube followed by the (optional) CD45/IgG1 control tube.

4. Display CD45 vs. side scatter and define a region, R1, around the CD45-positive cells (leucocytes), as a large rectangle (see *Figure 2A*).

5. Display an ungated plot of CD45 vs. CD34 (see *Figure 2B*) and define a second region,

Protocol 2 continued

R2, which includes all the CD34+ cells and somewhat fainter events. Leave this plot ungated and check that R1 is not excluding any CD34 cells expressing low levels of CD45, adjust R1 if necessary.

6. When you are satisfied with regions R1 and R2, begin to acquire the final data files. This will depend on the cytometer; you must count all events in R1, although it is not essential to list them all. If there is sufficient sample, obtain cells to list 3000 events in logical region (R1 and R2).

7. After acquiring each sample, record the number of cells you have acquired to get the 3000 R2 events, i.e. the total cells in R1.[b]

C. Analysing the data

1. Analyse the data sets from the 3000 collected events, but it is prudent to display the dot plots from the 50 000 events to visualize the cell scatter profile of the whole sample. Display side scatter vs. PE plots for the 3000 events (see *Figure 2C*). Place a region R3 around the population of CD34 positives, usually clearly seen as a cluster of cells with low side scatter.

2. Check the scatter profile of the CD34-positive, R3 events (see *Figure 2D*); these should fall in a region of low to medium forward scatter and low side scatter, characteristic of blast cells. Experience is essential at this stage; however, the scatter profile of lymphocytes (see *Figure 2A*, bright CD45 low side scatter) can be used as a guide. Adjust the region R3 if the cells have inappropriate scatter characteristics. Record at least 100 events for R3, this will result in CVs around 10%.

3. If you are using the optional control, repeat the measurement for this tube and subtract any background counts.[c]

[a] As small populations are under investigation, it is important to minimize any contamination. Between samples wipe the probe with a damp tissue and flush with clean sheath fluid.

[b] This will provide CVs better than 3% on all but very low samples. See the comment in Part B, step 2 if it is not possible to list 3000 R2 events.

[c] These should be very low compared with the sample counts, explaining why this is now considered an optional control. However, if the method is used to count very low samples, the effect of subtracting control counts will eventually become significant.

The International Society for Hematotherapy and Graft Engineering (ISHAGE) guidelines (6) are becoming increasingly popular as a method for defining the correct population of CD34+ cells. *Protocol 2* is somewhat simpler to follow and produces results that are not statistically different. It is recommended that you consult the ISHAGE guidelines in parallel with this protocol.

There have also been recent trends towards 'single platform' methods (7) where, by including a known concentration of beads with each sample, an absolute count can be obtained independent of the traditional method of using a count from another machine ('two platform' methods). Cytometers employing volumetric methods to derive an absolute count are also now appearing.

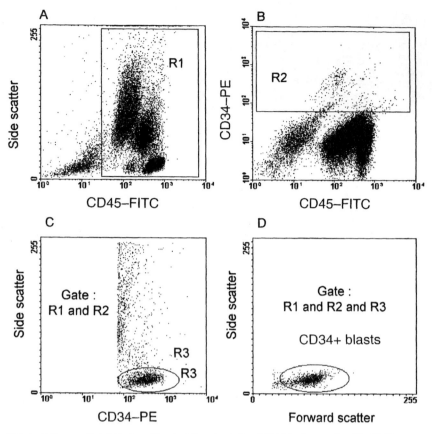

Figure 2 Sequential gating and back-gating to define CD34+ stem cells. (A) CD45
fluorescence is used to define leucocytes (R1). (B) CD34 fluorescence is used to define a
collection region (R2). (C) 3000 events collected in (R1 + R2); R3 defines cells of suitable
side scatter. (D) Scatter profile of the (R1 + R2 + R3) events; any debris or other unsuitable
events can be excluded from the final count.

4 Simultaneous detection of granulocyte- and lymphocyte-reactive antibodies

Detection and characterization of granulocyte-reactive antibodies in blood donors
and patients is of clinical importance because of their ability to cause febrile
transfusion reactions, transfusion-related lung injury, and immune neutropenia.
Detection of such antibodies has traditionally involved immunofluorescence
techniques (8, 9), but the granulocyte isolation procedures involved are often
complex and lengthy, resulting in considerable cell loss. Furthermore, when an
antibody reacting with granulocytes is detected, its true specificity may be
unclear since granulocytes express both HLA and granulocyte-specific antigens.
Resolution of specificity often requires further testing against lymphocytes from
the same donor, using a flow cytometric technique similar to that employed for
crossmatching in solid organ transplantation (see *Protocol 5*).

The problems of cell loss and discrimination between HLA and granulocyte-specific antibodies can be overcome by using a mixed cell population in an indirect immunofluorescence assay, and then using flow cytometry to identify and assess the individual populations. For example, lysed whole blood may be incubated with the serum under investigation, then washed and an FITC-conjugated anti-human Ig reagent added. The lymphocyte and granulocyte populations can then be identified by their light-scatter properties, and subsequently assessed for cell-bound antibodies. Consequently, granulocytes and lymphocytes are examined simultaneously, enabling differentiation between HLA and granulocyte antibodies in one test, thus avoiding complex cell isolation procedures.

A number of approaches to the technique have been described (10–12). The method given in *Protocol 3* is derived from a cell isolation procedure described by Hamblin *et al.* (13), and was originally described by Brough *et al.* (14). It has subsequently been used in a number of laboratories for the investigation of granulocyte allo- and autoantibodies (15).

Protocol 3

Simultaneous detection of granulocyte- and lymphocyte-reactive antibodies

Equipment and reagents

- EDTA-anticoagulated whole blood[a]
- Fixative: 0.16% (v/v) formaldehyde in PBS (1 in 250 dilution of stock 40% (w/v) solution)
- Lysing solution: 1.5 g Tris–HCl, 8.3 g NH$_4$Cl in 1000 ml H$_2$O, adjust pH to 7.4 with 10 M NaOH
- FCS
- 1% FCS in PBS

- U-bottom microplate
- Fluorescently conjugated anti-human Ig reagent (e.g. FITC–anti-IgG or anti-IgM), Fab, or F(ab')$_2$ preparation
- Control sera: inert negative-control sera and positive-control sera known to contain HLA and/or granulocyte-specific antibodies

A. Preparation of leucocyte suspension

1. Warm the lysing solution and fixative to 37 °C.
2. Add 4 ml of blood to a 50 ml conical-bottom tube and incubate at 37 °C for 5 min.
3. Add 4 ml of warm fixative and incubate at 37 °C for 4 min.[b]
4. Add 40 ml of warm lysing solution and incubate for 6 min at 37 °C, mix after 3 min.
5. Centrifuge the tube at 160 g for 5 min, decant the supernatant, and resuspend the cells by placing the tube on a vortex mixer.
6. Wash the cells twice, use 20 ml of 1% FCS in PBS and centrifuge for 5 min at 160 g each time.
7. After the second wash resuspend the cell pellet in 1 ml of 1% FCS in PBS.
8. Adjust the cell concentration to 5 × 10^6/ml using 1% FCS in PBS.

Protocol 3 continued

B. Incubation with sera

1. Prior to use, centrifuge all sera at 5000 g for 10 min to remove debris.
2. Add 50 μl of serum and 50 μl of cell suspension to the wells of a U-bottom micro-plate.
3. Incubate at 37 °C for 45 min.
4. Add 100 μl of 1% FCS in PBS to the wells containing cells and sera.
5. Centrifuge the microplate at 800 g for 5 min.[c]
6. Decant the supernatant and resuspend the cell pellets in 200 μl of 1% FCS in PBS.
7. Centrifuge the microplate at 800 g for 5 min.
8. Decant the supernatant and repeat the washing steps a further three times.

C. Incubation with fluorescently labelled anti-human Ig

1. Prepare an appropriate dilution of fluorescently conjugated anti-human Ig in 1% FCS in PBS.
2. Add 50 μl of the diluted anti-Ig reagent to the cell pellets remaining after decanting the last wash in step B8 (i.e. the fifth wash).
3. Mix and incubate in the dark at room temperature for 30 min.
4. Add 150 μl of 1% FCS in PBS and centrifuge at 800 g for 5 min.
5. Decant the supernatant, resuspend the cell pellet in 200 μl of 1% FCS in PBS.
6. Centrifuge at 800 g for 5 min.
7. Decant the supernatant, resuspend the cell pellet in 200 μl of 1% FCS in PBS.
8. Transfer the cells to appropriate tubes for flow cytometric analysis.

D. Measurement of cell-bound fluorescence

1. Process the samples on a flow cytometer, acquiring at least 20 000 cell events.
2. Draw analysis regions around the lymphocyte and granulocyte populations, and display fluorescence histograms for the two cell populations. Adjust the instrument settings so that the lymphocytes previously incubated with inert negative-control sera lie within the first decade, but away from the first channel.
3. Record the median fluorescence values for both cell populations and decide upon a negative/positive interpretation for test sera by comparing the negative control sera tested in parallel.

[a] Ideally blood should be tested on the day of collection, but it may be stored at room temperature overnight.

[b] To ensure that the contents of the tube are maintained at 37 °C, it is recommended that the water in a water bath reaches at least the 25 ml mark on the 50 ml tube.

[c] The centrifugation time may be varied, but should be as short as possible, whilst ensuring that all the cells are pelleted.

5 Detection of platelet-reactive antibodies

Platelet alloantibodies recognizing the human platelet antigens (HPA) cause a variety of clinical conditions such as neonatal alloimmune thrombocytopenia, post-transfusion purpura, and refractoriness to platelet transfusion, while auto-antibodies are the causative agents of autoimmune thrombocytopenias. Platelet-reactive antibodies are often detected using immunofluorescence techniques (16, 17), and the use of flow cytometric techniques appears to have improved the sensitivity of antibody detection methods (18).

Platelet alloantibodies are detected by incubating patient's sera with platelets isolated from donors with known HPA genotypes and subsequently detecting platelet-bound antibodies using fluorescently conjugated anti-human Ig reagents. *Protocol 4* is an example of a method, which may be used for detecting platelet alloantibodies. There are numerous variations on the technique; a web site maintained by the National Institute for Biological Standards and Control in the UK gives further protocols and information relating to platelet serology (http://www.nibsc.ac.uk/divisions/Haem/web2.html). Platelet autoantibodies are detected using similar protocols for isolating a patient's own platelets and subsequently testing with anti-human Ig reagents.

Protocol 4

Detection of platelet-reactive alloantibodies

Equipment and reagents

- EDTA- or citrate-anticoagulated whole blood
- 0.2 M EDTA. Titrate 340 ml of 0.2 M Na$_4$EDTA (83.2 g/l) against 500 ml of 0.2 M Na$_2$EDTA (74.5 g/l) to pH 6.8
- PBS/EDTA/BSA buffer: 0.01 M EDTA in PBS containing 0.25% (w/v) BSA

- U-bottom microplate.
- Fluorescently conjugated anti-human Ig reagent (e.g. anti-IgG, -IgA, or -IgM)
- Control sera: inert negative-control sera and positive-control sera known to contain HPA-specific antibodies

A. Platelet isolation and preparation

1. Centrifuge 10 ml of anticoagulated blood at 400 g for 10 min at room temperature.
2. Remove the plasma, which contains platelets, and transfer to a 10 ml centrifuge tube.
3. Fill the tube with PBS/EDTA/BSA buffer.[a]
4. Centrifuge the tube at 1500 g for 5 min.
5. Decant the plasma, mix the cell pellet, and resuspend in 10 ml of PBS/EDTA/BSA buffer.
6. Centrifuge the tube at 1500 g for 5 min.
7. Decant the supernatant, mix the cell pellet, and resuspend in 10 ml of PBS/EDTA/BSA buffer.
8. Centrifuge the tube at 1500 g for 5 min.

9. Decant the supernatant and resuspend the platelets to 5×10^7/ml with PBS/EDTA/BSA.

B. Incubation with sera

1. Prior to use, centrifuge all sera at 5000 g for 10 min to remove debris.
2. Add 50 μl of platelet suspension and 50 μl of serum to the wells of a U-bottom microplate.
3. Incubate at 37°C for 45 min.
4. Add 100 μl of PBS/EDTA/BSA to the wells containing platelets and sera.
5. Centrifuge the microplate at 1500 g for 5 min.
6. Decant the supernatant and resuspend the platelet pellets in 200 μl of PBS/EDTA/BSA.
7. Centrifuge the microplate at 1500 g for 5 min.
8. Decant the supernatant and repeat the washing steps a further three times.

C. Incubation with fluorescently labelled anti-human Ig

1. Prepare an appropriate dilution of fluorescently conjugated anti-human Ig in PBS/EDTA/BSA.
2. Add 50 μl of the diluted anti-Ig reagent to the cell pellets remaining after decanting the last wash in step B8 (i.e. the fifth wash).
3. Incubate at room temperature in the dark for 30 min.
4. Add 150 μl of PBS/EDTA/BSA and centrifuge at 1500 g for 5 min.
5. Decant the supernatant, resuspend the cell pellet in 200 μl of PBS/EDTA/BSA.
6. Centrifuge at 1500 g for 5 min.
7. Decant the supernatant, resuspend the cell pellet in 200 μl of PBS/EDTA/BSA.
8. Transfer the platelets to appropriate tubes for flow cytometric analysis.

D. Measurement of cell-bound fluorescence

1. Process the samples on a flow cytometer, identifying platelets by their scatter properties using logarithmic amplifier settings (see *Figure 1A*), and acquire about 5000 platelet events.[a]
2. Draw an analysis region around the platelets and display a fluorescence histogram.
3. Adjust the instrument settings so that the histogram for platelets previously incubated with inert negative-control sera lie within the first log decade, but away from the first channel.
4. Record the median fluorescence value for each platelet suspension and decide upon a negative/positive interpretation for each test sera by comparison with negative-control sera tested in parallel.

[a] All reagents should be freed from microparticles (e.g. by filtration) which may fall within the platelet region. In some instances, for example when working with platelets from thrombocytopenic patients, it may be advisable to use a two-colour technique, with a conjugated monoclonal antibody being used to positively identify platelets.

6 Application of crossmatching by flow cytometry to transplantation

Following the reports of Patel and Terasaki (19) and others, the introduction, in the early 1970s, of the cytotoxic crossmatch between donor cells and recipient serum made the hyperacute rejection of kidney allografts a comparatively rare event. However, early graft failure and/or accelerated rejection were still common and the possibility of a lack of sensitivity in the cytotoxic crossmatch was suggested. Over the next 10–15 years various improvements in the crossmatch methodology were proposed:

- enhancement of cytotoxicity by the addition of sublytic levels of anti-lymphocyte antibody;
- incubation at different temperatures and times;
- use of ^{51}Cr-labelled target cells;
- addition of dithiothreitol;
- separation of target cells into T and B cells.

In 1983, Garovoy *et al.* first described (20) the use of the flow cytometer to detect the presence of anti-donor antibody binding in the pre-transplant serum of kidney graft recipients to donor lymphocytes. Using only an FITC-conjugated anti-human IgG reagent, they showed not only that the flow cytometric method was up to 50 times more sensitive than the cytotoxicity test but also that it could distinguish between T and B lymphocyte binding. Most patients who showed positive antibody binding experienced graft failure, and it has been shown that flow cytometric crossmatching identifies a group of patients at risk from rejection. In the 1980s the technique was refined with the development of a two-colour technique (21). This involved the use of both an FITC-conjugated anti-IgG reagent and a PE-conjugated CD3 or CD20 to identify donor T or B cells, respectively. Important methodological considerations were examined over a 4-year period by the British Society of Histocompatibility and Immunogenetics flow group, and data showed that variations in technique can have significant effects on assay sensitivity and reproducibility (22). Over the last 5 years the flow crossmatch has become established as a routine clinical assay and has become a clinical prerequisite to transplantation.

Details of the crossmatch technique have been described by Bell *et al.* (23). Important in the methodology are adequate washing of the cell and serum mixture, checking the interaction between the FITC-conjugated polyclonal anti-human IgG and the monoclonal phenotyping antibodies, and the use of control AB sera. Many centres use panels of sera to define the negative result. Positivity of samples is best assessed in each centre by retrospective and prospective clinical trials.

Flow cytometry crossmatching has been applied to liver transplantation (24), while other developments have included the use of platelets or cell lines as the target cell source. The use of three-colour fluorescence has also been used to identify IgG binding to both T and B cells in the same sample (25).

Recently, flow cytometry has found clinical uses in antibody screening and its use may soon even replace the cytotoxic crossmatch as the 'gold standard' in pre-transplant crossmatching.

Protocol 5

The flow cytometric crossmatch

Equipment and reagents

- Ficoll–Hypaque (see Chapter 3, *Protocol 4*)
- PBS containing 1% FCS
- Test, and negative- positive-control sera
- 75 × 12 mm flow cytometer tubes
- Pre-titred anti-human IgG–FITC
- Anti-CD3–PE and anti-CD20–PE

A. Set up of crossmatch

1. Extract donor cells from the spleen or peripheral blood by Ficoll–Hypaque centrifugation (see Chapter 3, *Protocol 4*) and, after washing in PBS containing 1% FCS, adjust the lymphocyte concentration to 10^6 lymphocytes/ml.

2. Incubate 50 μl of cells (10^6 cells/ml) and 50 μl of test or negative- or positive-control sera in 75 × 12 mm flow cytometer tubes for 30 min in a 37°C water bath.

3. Remove the non-bound IgG in the serum by washing the cells three times with 4 ml of PBS/FCS for each wash.

4. Add 50 μl of pre-titred anti-human IgG–FITC and 5 μl of the anti-CD3–PE for T cells or anti-CD20–PE to each pellet, mix and incubate for 30 min at 4°C.

5. Remove remaining non-bound conjugates by washing the cells once with 4 ml PBS/FCS.

6. Resuspend the cells in 200 μl of PBS/FCS.

7. Analyse on an optimally set-up flow cytometer (see Chapter 5).

B. Data acquisition and analysis

1. Identify the lymphocyte population by its forward and side scatter light properties (see *Figures 3A and B*).

2. Analyse the lymphocytes by gating on region, R1 (as shown in *Figure 3*) and generating an FL2 (CD3–PE or CD20–PE) vs. FL1 (anti-IgG–FITC) dot plot from which the positively stained T or B lymphocytes can be identified (region R2 in *Figures 3C and D*).

3. Display a histogram of the fluorescence (FITC) (IgG) profile of the T or B cells to show the degree of anti-IgG binding (see *Figure 3E*).

4. Compare the average of the median, T-cell, peak-FITC fluorescence for the negative controls (comprising five panels of non transfused AB male sera) with the recipient test samples (run in triplicate). Calculate the difference between the two and express as a positive or negative result.

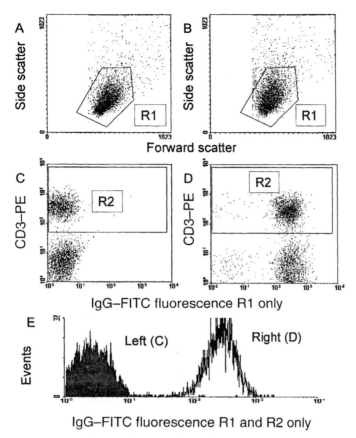

Figure 3 Analysis of IgG binding to T cells for a negative sample (A, C) and a positive sample (B, D). Region R1 (A, B) defines the lymphocytes by scatter profile and R2 (C, D) the T cells by CD3–PE fluorescence. The IgG–FITC fluorescence histograms of the gated populations are shown overlapped in E.

The definition of positivity needs to be established in each centre performing the crossmatch. It may be taken as test values being defined as positive when their median (of the three samples) is greater than the mean for that of the control tubes ± 2 SDs. Alternatively, positivity may be defined on the basis of channel-shift relative median fluorescence or on the basis of molecules of equivalent soluble fluorescein (MESF) (22).

7 Monitoring anti-thymocyte globulin (ATG)/anti-lymphocyte globulin (ALG) therapy in transplantation

Over the past few years, an increasingly wide range of drugs have been used to combat the rejection process associated with graft rejection. One of the main-stays of therapy, either as prophylaxis or for the treatment of rejection, has been the use of ALG. These polyclonal heterologous products act principally by reducing the level of peripheral blood lymphocytes. In 1963, Woodruff and Anderson (26) showed that anti-lymphocyte serum produced a lymphopenia that was mainly due to a reduction in the T-lymphocyte population. It has also been reported that the administration of ATG can increase the chances of infection. Currently, the majority of transplant centres administer ATG in an empirical fashion according to a fixed dose regimen. Due to the idiosyncratic response of individual patients, a fixed dose regimen may undersuppress some patients or result in oversuppression, which may, in turn, lead to acute renal allograft rejection or opportunistic infections such as cytomegalovirus.

There have been three methods described for monitoring the efficacy of ATG therapy. Serum IgG levels (27) have shown no correlation between serum levels and renal graft rejection. Cosimi *et al.* (28) found that circulating T-cell levels in patients receiving ATG were best detected using E-rosettes employing sheep erythrocytes. The conclusion was reached that the best therapy was achieved by reducing the level of circulating T lymphocytes to less than 10% of the patient's pre-transplant T lymphocyte count. This study was confirmed by Thomas *et al.* (29), who showed that the administration of ATG to keep the total circulating T-cell count less than 200 cells/μl reduced the incidence of acute allograft rejection and improved 1-year graft survival. The reproducibility of the E-rosette assay system was also questioned when flow cytometric measurement of monoclonal antibody-labelled T lymphocytes was introduced (30).

Flow cytometric analysis of monoclonal antibody-labelled T lymphocytes proved to be more accurate, more reproducible, and easier to perform (31) than the E-rosette assay. The efficacy of such treatment can be monitored using a threshold T-cell count of 50 T cells/μl (32). Flow cytometry provides an excellent, rapid and reproducible way of measuring such cells. This is achieved using three fluorescent markers. The lymphocytes are identified using side scatter (RALS) vs. anti-CD45 (PerCP) plots. The percentage of T cells can then be determined by the use of anti-CD3–FITC markers. The inclusion of anti-M3–PE removes monocytes from the gate. By spiking the labelled sample with fluorescent beads the number of T cells/μl can be calculated.

Protocol 6

T-cell monitoring in transplantation

Equipment and reagents

- EDTA-containing, blood collection tubes
- Anti-CD45–PerCP(or ECD)
- Anti-CD14–PE
- IgG–FITC or CD3–FITC
- FACS lysing solution (Becton Dickinson)
- Flowcount beads (Beckman–Coulter)

A. Cell preparation

1. Collect the blood into EDTA containing tubes. Set up two tests.

Protocol 6 continued

2. Add 100 μl of mixed anticoagulated blood to each of the following two tubes and

   either:

   (a) 5 μl anti-CD45–PerCP(or ECD) + 5 μl anti-CD14–PE + 5 μl IgG–FITC isotype control (of the same isotype as the CD3 antibody)

   or

   (b) 45 μl anti-CD45–PerCP(or ECD) + 5 μl anti-CD14–PE + 5 μl CD3–FITC.

3. Vortex the tubes at low speed for 3 sec and then incubate at room temperature in the dark for 30 min.

4. Add 0.5 ml of 1 × FACS lysing solution to each tube and vortex immediately.

5. Incubate for 10 min at room temperature in the dark; add 100 μl of Flowcount beads before analysing on a flow cytometer.

B. Data acquisition and analysis

1. Construct dot plots of side scatter against anti-CD45–PerCP (see *Figure 4A*) and draw a region, R1, to define the lymphocytes. Draw another region round the polyfluorescent bead population.

2. Use quadrant statistics to calculate a CD14–PE vs. CD45–FITC dot plot to determine lymphocyte purity in the gate.

3. Determine the number of lymphocytes expressing CD3–FITC after using the isotype control to set a 3% marker (see *Figure 4B*). Calculate the absolute count of lymphocytes/μl, i.e.

   lymphocytes counted (R1) × beads per μl/beads counted (R2)

4. From this, calculate the number of T cells from the percentage of positive cells in the second tube relative to that of the isotype control set at 3%.

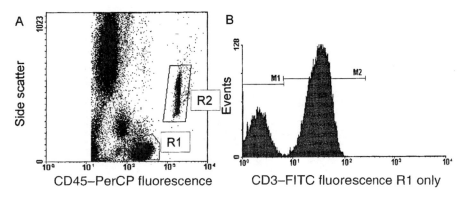

Figure 4 (A) CD45–PerCP and bead fluorescence plotted against side scatter. Region R1 defines the lymphocytes and R2 the beads. (B) CD3 fluorescence of the lymphocytes in region R1, the positive T cells are counted by marker M2, the markers having been set on a control sample.

113

8 Measurement of cell-mediated cytotoxicity by flow cytometry

Cell-mediated cytotoxicity can be measured by a dye-exclusion assay (33, 34). The advantages of this are the lack of requirement for radioisotopes and the fact that killing can be measured on a 'per cell basis', irrespective of whether the mechanism is necrotic or apoptotic. This means that one can calculate the precise percentage of cells killed by the effector cells directly, rather than in-direct extrapolation as is the case in ^{51}Cr-release assays. Traditional isotope-release assays require that all target cells label uniformly with the isotope. This is commonly assumed to be the case when using cell lines as targets, but it may not be true of clinically relevant targets such as leukaemic blasts.

The cytotoxicity assay described here can be used with cell lines or clinical isolates; the procedure is identical. It uses PKH26 (Sigma) to label the target cells. The labelled cells show high levels of orange fluorescence (see *Figure 5A*)

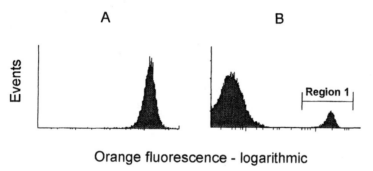

Figure 5 Orange fluorescence from PKH26-labelled cells (A) and when mixed with non-labelled effector cells (B, region 1).

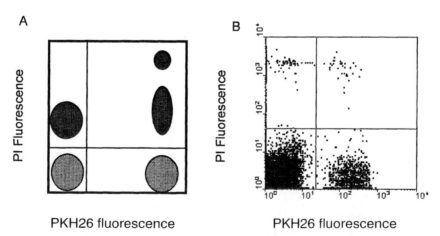

Figure 6 (A) Theoretical and (B) practical bivariate fluorescence distributions from a mixture of effector and target cells. Target cells appear in the right-hand quadrants and apoptotic/dead cells in the top quadrants.

and can easily be resolved from non -labelled effector cells (region 1 in *Figure 5B*). The membrane dye is very stable and there is no transfer to non-labelled effector cells during assay periods of up to 48 h.

In *Figure 6B* the PI+ve target cells have been analysed as a single group, although it is apparent that they fall into two distinct subsets. The PI weakly positive cells shown in *Figure 6B* are early apoptotic cells, where the PI has passed into the cytoplasm but not intercalated with the DNA since the nuclear membrane remains intact.

Protocol 7

Measurement of cell-mediated cytotoxicity

Equipment and reagents

- Equipment and reagents for density-gradient separation (see Chapter 3, *Protocol 4*)
- PKH26 cell-labelling buffer (Sigma)
- FCS
- Tissue culture medium supplemented with 10% FCS
- 1 μg/ml propidium iodide (Sigma or Molecular Probes) in PBS

A. Labelling the target cells

1. Isolate target cells by density-gradient separation (see Chapter 3, *Protocol 4*).
2. Suspend the target cells in PKH cell-labelling buffer (Sigma) at a concentration of up to 10^7/ml.
3. Mix the cells with an equal volume of 4 mM PKH26 for 2 min at 21°C in a 1.5 ml Eppendorf tube.
4. Stop the reaction by adding an equal volume of neat fetal calf serum (FCS) for 1 min.
5. Wash the cells twice in tissue culture medium supplemented with 10% FCS. Washing can be performed in a microcentrifuge operated at 7000 r.p.m. for 20–30 sec.
6. Resuspend the washed target cells in tissue culture medium supplemented with serum.

B. Mixing with the effector cells at the appropriate ratio

1. Incubate the effector cells with the labelled target cells at ratios of 10:1 to 50:1 for up to 48 h, although 4- or 20-h assays are more usual.
2. After the incubation period, resuspend the cells in a solution of 1 μg/ml propidium iodide in PBS.

C. Analysis by flow cytometry

1. Set the PMT voltages and fluorescence compensation settings using labelled target cells alone with and without PI/PBS as described in Part A, steps 1–4.
2. Start by running the labelled target cells without PI. Switch off all fluorescent

compensation. Acquire PKH26 fluorescence on FL2 (~560 nm) and set the PMT voltage such that the labelled cells fall within the third log decade on a four-log instrument.

3. Display a dot plot of FL2 vs. FL3 and adjust the FL3–FL2 compensation such that the PKH26+ve cells fall in the first 50 channels of the FL3 display.

4. Add the PI/PBS to the sample of target cells and re-run. Adjust the FL3 PMT voltage such that a negative population falls within the first log decade of the FL3 display. Increase the FL3 PMT to ensure that this is the negative population (i.e. there is not a second population below the first) and then return the PMT to its appropriate setting.

5. Run a tube containing effector cells alone suspended in PI/PBS. Adjust the FL2–FL3 compensation such that any PI+ve effector cells fall within the first log decade of the FL2 PMT. Note that after successful PMT and compensation adjustment the display of effector and target cell combinations should appear as in *Figure 6A*, although they are more likely to appear as in *Figure 6B*.

6. Acquire at least 10 000 target cells with 1024 channel resolution after electronic gating on PKH26 fluorescence.

7. Determine the mean proportion of PI+ve cells from the triplicate samples.

8. Determine the background target cell death from cells incubated in the absence of effector cells.

9. Report cell-mediated cytotoxicity as the percentage killing over background cell death averaged from the three samples: mean (% PI+ve in test – % background death).

9 Monitoring cell activation in response to antigens or mitogens

The early activation antigen CD69 is an excellent marker of lymphocyte activation *in vitro* and can be used to monitor immune responsiveness (35). Indeed, we have shown that depletion of alloreactive cells in a mixed lymphocyte reaction (MLR) by virtue of CD69 expression abrogates the specific alloreaction from a subsequent MLR, whilst retaining lymphocytes with a capacity to respond to irrelevant third-party alloantigens (36).

For analysing cellular responses to soluble antigens or mitogens it is sufficient to incubate the responder cells with the stimulus in culture medium for 4–48 hours. The time course is dependent upon the stimulus (37) (see *Figures 7A and B*). For the analysis of specific cell populations within a heterogeneous cell mixture it is often best to label with multiple fluorochrome-conjugated monoclonal antibodies (mAbs). This is facilitated by the availability of both FITC- and PE-conjugated anti-CD69 mAbs, both of which give a sufficient signal-to-noise ratio to allow discrimination from negative cells by cluster analysis.

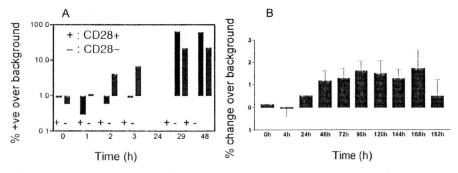

Figure 7 (A) Time course of phytohaemagglutinin-induced CD69 expression on CD8+ cells. (B) Temporal expression of CD69 on CD3+ T cells in a mixed lymphocyte reaction.

The ability to study subsets of cells within heterogeneous starting populations is very useful since it allows *in-vitro* analysis of cellular reactions without prior cell subset selection.

For analysing cell activation in response to a cellular stimulus, such as in an MLR, one can use the PKH26 membrane dye to label the stimulators and thus exclude them from the analysis of the responder cell population. *Figure 8* is a dot plot in which CD69–FITC signals are arrayed against CD56–PE events in an MLR, in which the stimulator cells have been labelled with PKH26. The stimulator cells are clearly distinguishable as the FITC (FL1)+/PE (FL2)++ cluster of cells. CD56+ cells in the responder population are readily identifiable, and the proportion activating in response to the stimulus can be calculated.

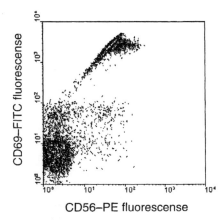

CD56–PE fluorescense

Figure 8 Bivariate fluorescence distributions resulting from a mixed lymphocyte reaction. The PKH26-labelled stimulator cells appear as the top population. The lower two decades represent the CD56 vs. CD69 distribution of the responder cells.

10 Measuring intracellular cytokines by flow cytometry

The immune response mechanism involves complex communication between cells. Signalling is mediated both by cell–cell interactions and by secretion and reception of soluble factors such as cytokines. At the start of the immune response, activated antigen-presenting cells (such as dendritic cells, DC, and monocytes) secrete cytokines (38). The different cytokines released affect different lymphocyte subpopulations, causing them, in turn, to secrete their own cytokines. The pattern of cytokine release depends on the type of immune response, and is regulated by both positive and negative feedback loops. A simplistic, but useful, description of lymphocyte cytokine secretion patterns is the T_{H1}/T_{H2} paradigm (39). The T_{H1} cytokine pattern involves pro-inflammatory cytokines such as interferon-gamma (IFN-γ), produced in response to monocyte interleukin (IL)-12, and the T_{H2} pattern involves cytokines such as IL-4 and IL-5 important for allergic responses. In many diseases, there are abnormalities of the pattern of cytokine release involving monocytes, DC, and lymphocytes. An important example is the susceptibility to mycobacterial infection resulting from defects in the circuit involving monocyte IL-12 and subsequent lymphocyte IFN-γ and tumour necrosis factor-α (TNF-α) production (40–42).

Early studies measured cytokines secreted by stimulated cells in culture, often by ELISA of culture supernatants. Although these gave valuable information, they were unable to distinguish which cell type was making which cytokine. The use of cell clones or cell lines produced more insights, but still did not allow the effects of complex cellular interactions to be studied. *In situ* hybridization methods for the detection of cytokine mRNA partially overcame this problem, but had their own disadvantages, especially since there is poor correlation between cytokine mRNA levels and functional protein production. Indirect immunofluorescence techniques overcame the problem of identifying cytokine protein within individual cells, but were limited by an inability to delineate lymphocyte subset markers and cytokine production simultaneously.

The development of flow cytometric methods to detect newly synthesized cytokines within cells was a major step forward (43), allowing the cytokines to be identified within defined cell subpopulations and for rare cell populations to be studied. The principle of the method is to measure the capacity of the cells to make cytokines. Usually, cells in the circulation do not synthesize cytokines. A short *in vitro* stimulation using a potent stimulus (e.g. a phorbol ester and ionomycin) induces the production of cytokines reflecting the current state of the cells. After stimulation, differences in the production of cytokines between different samples (e.g. from patients or normal donors) measure differences in the ability to make cytokines at the time of donation, and thus ultimately reflect differences in the ability to make cytokines *in vivo*. To detect each cytokine by flow cytometry after the *in vitro* stimulation, the cells are fixed and permeabilized to allow the relevant detecting anti-cytokine antibodies to pass through the cell membrane and access the cytokine. The sensitivity of the method is enhanced by the use of reagents, such as monensin or brefeldin A, that retain the newly synthesized cytokines within the cells by blocking their passage through the Golgi apparatus (44).

Technical advances to the method have been many. The establishment of a range of different fluorochromes directly conjugated to monoclonal antibodies has allowed three- and four-colour cytometry to be done simply. In initial studies, cells were permeabilized using saponin, but now quality-controlled commercial reagents are usually used for fixation and permeabilization (e.g. Leuco-Perm solutions A and B, Serotec). In most instances, it is possible to add all the monoclonal antibodies at the same time, including those used to identify cell subpopulations by surface marking and those detecting intracellular cytokines. However, there are some antibodies, e.g. anti-CD14 and anti-CD56, that have to be used before the fixation and permeabilization steps, since the antigen may undergo conformational changes and hence stain poorly.

For the intracellular cytokine work, we have used both whole blood and mononuclear cell cultures (45, 46). The whole-blood method we developed requires small volumes of blood, is rapid, and robust (47).

Protocol 8

Staining for intracellular cytokines (whole blood)

Equipment and reagents

- Lithium heparin
- RPMI-1640 (Life Technologies)
- Monensin (Calbiochem)
- Phorbol myristate acetate (PMA, Calbiochem)
- Ionomycin (Calbiochem)
- Optilyse C (Beckman–Coulter)

- LeucoPerm (reagents A and B, Serotec)
- PBS/Az: 0.1% Na azide in PBS
- PBS/Az/BSA: PBS/Az plus 1% BSA (Sigma)
- 0.5% paraformaldehyde in PBS
- Flowcount beads (Beckman–Coulter)

Method

1. Take blood into lithium heparin (1–2 ml); note that EDTA or citrate is not suitable.
2. For each culture tube, dilute 250 μl blood (1 in 3) with 500 μl of RPMI-1640 medium.
3. Set up stimulated and unstimulated cultures (with and without 10 ng/ml PMA and 2 μmol/l ionomycin); include 3 μmol/l monensin in all cultures to retain cytokine within the cells.
4. Culture for 2 h at 37°C.
5. If using antibodies to surface markers that are sensitive to inactivation by permeabilization (e.g. anti-CD14 or anti-CD56), stain with all anti-cell surface antibodies at this time, if there is no such problem wait until step 12. Add directly conjugated antibodies to the culture and incubate in the dark at room temperature for 15 min.
6. Lyse red cells for 15 min with appropriate lysing solution (e.g. if employing a Coulter EPICS cytometer, use 2 ml of Optilyse C for the original 250 μl of blood).
7. Wash in PBS/Az/BSA, centrifuge at 250 g for 5 min at room temperature and discard the supernatant.
8. Fix in 250 μl LeucoPerm reagent A for 15 min.[a]
9. Wash in PBS/Az/BSA, centrifuge at 250 g for 5 min at room temperature and discard the supernatant.

Protocol 8 continued

10. Permeabilize in 250 μl LeucoPerm reagent A for 15 min.[a,b]

11. Add 50 μl aliquots of cells in solution B to tubes already containing the directly conjugated antibodies against surface antigens (unless already done at step 6) and the directly conjugated antibodies against the relevant cytokines.[b]

12. Incubate in the dark at room temperature for 30 min.[b]

13. Wash in PBS/Az by centrifuging at 250 g for 5 min at room temperature.

14. Store with paraformaldehyde (0.5% in PBS) at 4°C until acquisition within 24 h.

15. Just before acquisition, add a constant aliquot of Flowcount beads to each tube— the aliquot volume is equal to the blood volume (i.e. 50 μl in this example).

[a] Samples must be well drained before adding the LeucoPerm reagent A or B.

[b] It is possible to combine steps 11, 12, and 13 by adding the antibodies directly to the 30-min permeabilization step in tubes with 50 μl cell aliquots, but extra reagent B may be needed to avoid lack of permeabilization due to its dilution.

Protocol 9

Acquisition and analysis for intracellular cytokines

Equipment and reagents

- Four-colour flow cytometer[a] (e.g. single laser, four-colour Coulter EPICS-MCL flow cytometer)

- Flowcount beads (Beckman–Coulter)

Method

1. *Absolute counts.* Add an aliquot of Flowcount beads to each tube just before acquisition to allow the absolute numbers of all cell populations to be determined.

2. *Order of tubes.* Analyse the tubes in pairs with the same anti-cytokine antibody (and the other antibodies) in each, with the appropriate unstimulated control (e.g. monensin only) preceding the stimulated sample.

3. *Boolean combinations of regions to define gates.* As an example, using the first antibody combination mentioned in footnote *b* it is possible to measure the IFN-γ positive cells in the following populations: CD3+, CD4+ (by CD3+CD8−), CD8+, and the CD28 positive and negative populations of each of them. Do this by defining regions on dot plots and combine these regions into gates.

4. *Definition of cytokine-positive cells.* Note that cytokine-positive cells are those clearly brighter than the negative cells.[c]

[a] The use of four colours increases, by an order of magnitude, the power of studying cytokines in different cell populations. In some instances information can not be obtained with less than four colours.[b] We use a single laser four colour Coulter EPICS-MCL flow cytometer.

Protocol 9 continued

[b] Examples of 4-antibody combinations used are anti-IFN-γ/CD28/CD8/CD3, TNFα/CD56/CD8/CD3 and CD14/IL12/CD3/HLADR (Fluorochrome sequence FITC/PE/ECD/PeCy5).

[c] Small shifts in the position of the negative curve itself do not represent unequivocal cytokine-positive cells. Note that it is not usually valid to use an irrelevant antibody for the negative control for an anti-intracellular cytokine antibody. This is because most antibodies exhibit some non-specific binding, especially when intracellular epitopes and binding sites are available after permeabilization. Since this non-specific binding varies for different antibodies it is safer to use the anti-cytokine antibody as its own control with the cells in the unstimulated state. If there were some intrinsic expression of intracellular cytokine a more suitable negative control would be an excess of added free cytokine.

We will illustrate the usefulness of measuring intracellular cytokines with an application from our own work in which we have studied the primary immunodeficiency disease, common variable immunodeficiency (CVID).

In CVID, B lymphocytes are present but they fail to make antibodies and the patients have low levels of immunoglobulins of all isotypes (48). There may be

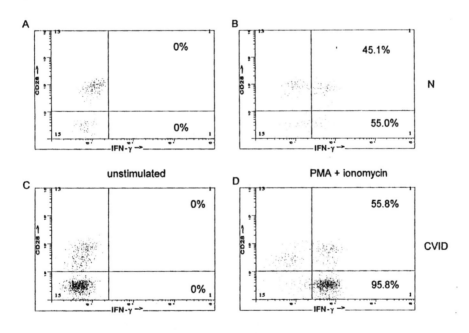

IFN-γ (%) in CD28 subsets of CD8+ T cells

Figure 9 Intracellular IFN-γ measured in CD8+ lymphocytes from a single normal donor (N) (top) and a single CVID patient (bottom), before (left) and after (right) stimulation *in vitro* of a whole blood sample with PMA and ionomycin for 2 h. There is no intrinsic IFN-γ before stimulation. After stimulation, the percentage expression of IFN-γ is greater in the CD28-negative cells in both donors, but the CVID patient has cells with more IFN-γ. Note the reversal of the CD28-positive cells in the two donors; there are relatively more in the normal donor and relatively less in the CVID patient. (Data from Cambronero *et al.*, (50)).

defects in the B cells themselves, but it is clear that a major problem in this disease is a failure of the T lymphocytes to provide the required help for the B lymphocytes to mature and produce antibodies. We have shown by limiting-dilution techniques that CVID T cells do not produce antigen-specific memory T cells (49).

Figure 9 shows examples of the measurement of intracellular IFN-γ in sub-populations of CD8+ T cells, the CD28 positive and negative subsets. It shows the percentage expression of IFN-γ in these two subsets in a single normal donor and a single CVID patient before and after stimulation (Cambronero *et al.*, (50)). There is no intrinsic expression of IFN-γ without stimulation, but after stimul-ation there is more IFN-γ in the CD28– cells and also more in the CVID patient's cells than in the normal's. The same was true in CD4+ cells (not shown). This data was significant when 12 patients were compared with 12 normal donors and confirmed our previous data (45, 46). We have also shown that the increase in lymphocyte IFN-γ in CVID correlates with an increase in monocyte IL-12, the cytokine known to be a potent up-regulator of IFN-γ. Thus it is clear that CVID is a disease polarized in relation to cytokine profiles towards a T_{H1} pattern. Interestingly, we have been able to demonstrate that many examples of an increased percentage expression of cytokine-positive cells is not due to an increase in the absolute number of cytokine-positive cells, but rather to a depletion of cells negative for the cytokine.

Note

Sections 2 and 3 were written by T. Hoy; Sections 4 and 5 by S. Garner; Sections 6 and 7 by B. K. Shenton and A. E. Bell; Sections 8 and 9 by M. Lowdell; and Section 10 by J. Farrant, M. North and C. Sewell.

References

1. Lee, L. G., Chen, C. H., and Chiu, L. A. (1986). *Cytometry*, **7**, 508.
2. Hoy, T. (1994). In *Flow cytometry: clinical applications* (ed. M. G. Macey), p. 192. Blackwell Scientific, Oxford.
3. Davis, B. H. and Biglow, N. (1989). *Arch. Pathol. Lab. Med.*, **113**, 684.
4. Holyoake, T. L. and Alcorn, M. J. (1994). *Blood Rev.*, **8**, 113.
5. Sutherland, D. R., Keating, A., Nayar, R., Anania, S., and Stewart, A. K. (1994). *Exp. Haematol.*, **22**, 1003.
6. Sutherland, D. R., Anderson, L., Keeney, M., Nayar, R., and Chin-Yee, I. (1996). *J. Hematotherapy*, **5**, 213.
7. Keeney, M., Chin-Yee, I., Weir, K., Popma, J., Nayar, R., and Sutherland, D. R. (1998). *Cytometry*, **34**, 61.
8. Verheugt, F. W., von dem Borne, A. E., Decary, F., and Engelfriet, C. P. (1977). *Br. J. Haematol.*, **36**, 533.
9. Verheugt, F. W., von dem Borne, A. E., van Noord Bokhorst, J. C., and Engelfriet, C. P. (1978). *Br. J. Haematol.*, **39**, 339.
10. Robinson, J. P., Duque, R. E., Boxer, L. A., Ward, P. A., and Hudson, J. L. (1987). *Diagn. Clin. Immunol.*, **5**, 163.
11. Veys, P. A., Gutteridge, C. N., Macey, M., Ord, J., and Newland, A. C. (1989). *Vox. Sang.*, **56**, 42.

12. Sintnicolaas, K., de Vries, W., van der Linden, R., Gratama, J. W., and Bolhuis, R. L. (1991). *J. Immunol. Meth.*, **142**, 215.

13. Hamblin, A., Taylor, M., Bernhagen, J., Shakoor, Z., Mayall, S., Noble, G., and McCarthy, D. (1992). *J. Immunol. Meth.*, **146**, 219.

14. Brough, S., Garner, S. F., and Lubenko, A. (1994). *Platelets*, **5**, 223.

15. Lubenko, A. and Wilson, S. (1996). *Immunohaematology*, **12**, 164.

16. von dem Borne, A. E., Verheugt, F. W., Oosterhof, F., von Riesz, E., de la Riviere, A. B., and Engelfriet, C. P. (1978). *Br. J. Haematol.* **39**, 195.

17. Rosenfeld, C. S. and Bodensteiner, D. C. (1986). *Am. J. Clin. Pathol.*, **85**, 207.

18. Allen, D. L., Chapman, J., Phillips, P. K., and Ouwehand, W. H. (1994). *Transfus. Med.*, **4**, 157.

19. Patel, R. and Terasaki, P. I. (1969). *N. Engl. J. Med.*, **280**, 735.

20. Garovoy, M. R., Rheinschmidt, M. A., Bigos, M., Perkins, H., Colombe, B., Feduska, N., and Salvatierra, O. (1983). *Transplant. Proc.*, **15**, 1939.

21. Cook, D. J., Terasaki, P. I., Iwahi, G. Y., Terashita, G. Y., and Lau, M. (1987). *Clin. Transplant.*, **1**, 253.

22. Shenton, B. K., Bell, A. E., Harmer, A. W., Boyce, M., Briggs, D., Cavanagh, G., Culkin, J., van Damm, M. G., Evans, P. R., Haynes, P., Henderson, N., Horsburgh, T., Martin, S., Preece, K., Reynolds, W., Robson, A., Sutton, M., Waters, D., Younie, M., and Garner, S. (1997). *Transplant. Proc.*, **29**, 1454.

23. Bell, A., Shenton, B. K., and Garner, S. (1998). *Proc. R. Soc. Med.*, **33**, 219.

24. Ogura, K., Terasaki, P. I., Koyama, H., Chia, J., Imagawa, D. K., and Busuttil, R. W. (1994). *Clin. Transplant*, **8**, 111.

25. Robson, A. and Martin, S. (1996). *Transplant. Immunol.*, **4**, 203

26. Woodruff, M. F. A. and Anderson, N. F. (1963). *Nature*, **200**, 702.

27. McAlack, R. F., Stern, S., Beizer, R., Sklar, L., and Bannett, A. (1979). *Transplant. Proc.*, **11**, 1431.

28. Cosimi, A. B., Wortis, H. H., Delmonico, F. L., and Russell, P. S. (1976). *Surgery*, **80**, 155.

29. Thomas, F., Lee, H. M., Wolf, J. S., Mendez-Picon, G., and Thomas, J. (1976). *Surgery*, **79**, 408.

30. Cosimi, A. B., Colvin, R. B., Burton, R. C., Rubin, R. H., Goldstein, G., Kung, P. C., Hansen, W. P., Delmonico, F. L., and Russell, P. S. (1981). *N. Engl. J. Med.*, **305**, 308.

31. Ganghoff, O., Gross, U., Whitley, T. K., Roebuck, D., and Thomas J. M. (1985). *Transplant. Proc.*, **17**, 640.

32. Shenton, B. K., White, M. D., Bell, A. E., Clark, K., Rigg, K. M., Forsythe, J. L. R., Proud, G., and Taylor, R. M. R. (1994). *Transplant. Proc.*, **26**, 3177.

33. Hatam, L., Schuval, S., and Bonagura, V. R. (1994). *Cytometry*, **16**, 59.

34. Lowdell, M. W., Ray, N., Craston, R., Corbett, T., Deane, M., and Prentice, H. G. (1997). *Bone Marrow Transplant.*, **19**, 891.

35. Maino, V. C., Suni, M. A., and Ruitenberg, J. J. (1995). *Cytometry*, **20**, 127.

36. Koh, M. B. C., Prentice, H. G., and Lowdell, M. W. (1999). *Bone Marrow Transplant.* (In press.)

37. Craston, R., Koh, M., McDernott, A., Ray, N., Prentice, H. G., and Lowdell, M. W. (1997). *J. Immunol. Meth.*, **209**, 37.

38. Trinchieri, G. (1997). *Curr. Opin. Immunol.*, **9**, 17.

39. Romagnani, S. (1996). *Clin. Immunol. Immunopathol.*, **80**, 225.

40. Cooper, A. M., Roberts, A. D., Rhoades, E. R., Callahan, J. E., Getzy, D. M., and Orme, I. M. (1995). *Immunology*, **84**, 423.

41. Frucht, D. M. and Holland, S. M. (1996). *J. Immunol.*, **157**, 411.

42. Jouanguy, E., Lamhamedi-Cherradi, S., Altare, F., Fondaneche, M. C., Tuerlinckx, D., Balnche, S., Emile, J-F., Gaillard, J-L., Schreiber, R., Levin, M., Fisher, A., Hivroz, C., and Casanova, J-L. (1997). *J. Clin. Invest.*, **100**, 2658.

43. Sander, B., Andersson, J., and Andersson, U. (1991). *Immunol. Rev.*, **119**, 65.

44. Jung, T., Schauer, U., Heusser, C., Neumann, C., and Rieger, C. (1993). *J. Immunol. Meth.*, **159**, 197.

45. North, M. E., Ivory, K., Funauchi, M., Webster, A. D. B., Lane, A. C., and Farrant, J. (1996). *Clin. Exp. Immunol.*, **105**, 517.

46. North, M. E., Webster, A. D. B., and Farrant, J. (1998). *Clin. Exp. Immunol.*, **111**, 70.

47. Sewell, W. A. C., North, M. E., Webster, A. D. B., and Farrant, J. (1997). *J. Immunol. Meth.*, **209**, 67.

48. WHO Scientific Group (1995). *Clin. Exp. Immunol.*, **99**, 1.

49. Kondratenko, I., Amlot, P. L., Webster, A. D., and Farrant, J. (1997). *Clin. Exp. Immunol.*, **108**, 9.

50. Cambronero, R., Sewell, W. A. C., North, M. E., Webster, A. D. B., and Farrant, J. (2000). *J. Immunol.*, **164**, 488.

Chapter 8

Quality assurance in the clinical laboratory

Jan W. Gratama* and David Barnett[†]

*Department of Clinical and Tumor Immunology, Daniel den Hoed Cancer Center, PO box 5201, 3008 AE Rotterdam, The Netherlands
[†]Department of Haematology, Royal Hallamshire Hospital, Sheffield, UK

1 Introduction

Each clinical diagnostic laboratory must have a strategy to assure the quality of its products. The focus in this chapter is on flow cytometric immunophenotyping assays; cases in point are:

for *quantitative assays*:

(a) enumeration of lymphocyte subsets, and

(b) enumeration of CD34+ haematopoietic progenitor cells (HPC); and

for *qualitative assays*:

(c) leukaemia/lymphoma immunophenotyping, and

(d) screening for the HLA-B27 antigen.

Quality assurance strategy includes:

- the monitoring, evaluation, and implementation of laboratory procedures for the pre-analytical, analytical, and post-analytical testing to ensure that correct results are obtained and reported;

- internal quality-control (QC) procedures and participation in an external quality assessment or proficiency testing programme.

As a result, reliable discrimination can be made between biological variations (i.e. disease processes) and process variations (i.e. due to the analytical system).

2 Reference materials for calibration

Assay calibration is an important component of internal QC and constitutes the evaluation (and adjustment) of the test system to provide a known relationship between the measurement response and the value of the substance being measured by the test procedure (1). Assay calibration makes use of reference

125

Table 1 Reference materials and predicate assays for four, clinical diagnostic, flow cytometric assays

Clinical diagnostic, flow cytometric assay	Possible source of reference material	Predicated method ('gold standard')
Quantitative assays		
Lymphocyte subset assay enumeration	Peripheral blood	'Single-platform' 3- or 4-colour (2, 3)
CD34+ HPC enumeration	Peripheral blood, apheresis product, or cord blood	'Single-platform' assay based on counterstaining of CD34 with CD45 (4)
Qualitative assays		
Leukaemia/lymphoma immunophenotyping	Peripheral blood[a]	Not defined
Screening for HLA-B27 antigen	Peripheral blood	DNA typing (5), complement-dependent cytotoxicity (6)

[a]Containing abnormal cells. Bone marrow would have been the ideal source of reference material, as most patient specimens are bone-marrow aspirates, but this is not available for QC purposes for ethical reasons.

materials with a known value or characteristic assigned by a predicated method ('gold standard'). Such reference materials are also used for external quality assurance surveys (EQAS). Of critical importance is that the reference material is stable over time (at least several weeks, preferably months) and has interassay properties comparable with those of patient specimens—i.e. no components in the reference material, other than those to be analysed, should influence the measurement and thereby the parameters to be quantified. In other words, the reference material should have minimal matrix effects. *Table 1* presents an overview of possible sources of reference materials for the four, clinical diagnostic, flow cytometric assays mentioned above.

As fresh blood specimens will already show some form of decay within hours, it is obvious that stabilization is necessary. Cryopreservation is still being used in one leukaemia/lymphoma immunophenotyping EQAS (7) and has the disadvantage that (a) density-gradient enrichment of mononuclear cells must be performed before freezing, and (b) significant artefacts occur after thawing (e.g. loss of cells, increased proportions of dead cells). Lyophilized lymphocytes (Cytotrol; Beckman–Coulter) also are quite distinct from whole blood specimens. The same manufacturer has engineered the human KG1a cell line to express similar levels of CD34 and CD45 as normal haematopoietic progenitor cells (HPC); a suspension of these cells in a predefined concentration (Stem-Trol) can be used as a procedure control for CD34+ HPC enumeration after spiking into fresh blood specimens. An alternative procedure for stabilizing blood specimens with minimal matrix effects has been used successfully by UK NEQAS for Leucocyte Immunophenotyping for lymphocyte immunophenotyping (8), CD34+ HPC enumeration (9), and leukaemia/lymphoma immunophenotyping. Reference material for lymphocyte immunophenotyping stabilized according to this procedure has been available as Ortho AbsoluteControl (Ortho-Clinical Diagnostics). Similar 'whole blood-like' products for lymphocyte immunophenotyping,

CD34+ HPC enumeration, and screening for HLA-B27 are available from Bio-Ergonomics, R & D Systems, and Streck Laboratories. *Protocol 1* describes a procedure for the 'do-it-yourself' stabilization of fresh whole blood samples using the StabilCyte cell-stabilization kit (Bio-Ergonomics). Matrix effects should be minimal up to 1 week of storage at 4°C; after that, red cells may become resistant to lysis. The stabilized leucocytes are claimed to retain their light-scatter scatter and immunophenotypic characteristics native to the cell type for 3–9 months with appropriate storage.

Protocol 1

Stabilization of whole blood samples

Reagents

- The StabilCyte cell-stabilization kit consists of two reagents: stabilization buffer A and stabilization buffer B. Store both buffers at 4°C until the expiration date (i.e. ≤12 months). Use the reagents, as is, without further preparation, but warm them to room temperature prior to use.
- PBS containing 10% serum (human, bovine, or fetal), pH 7.2

Method

1. Perform a white blood cell (WBC) count. If the WBC count exceeds 2×10^7/ml, dilute the sample using PBS containing 10% serum to achieve a WBC count of $1-2 \times 10^7$/ml.
2. Add an equal volume of buffer A to the volume of blood sample needed. Mix well.
3. Place the sample on a rocking platform or rotator at low speed and incubate for 2 h at room temperature.
4. After incubation, centrifuge the cell suspension at low speed (e.g. 450 g for 5–10 min), whilst still ensuring the full recovery of leukocytes. This step serves to remove dead cells and debris that may in the long run reduce the quality of the stabilized leukocytes.
5. Aspirate as much as possible of the supernatant without disturbing the cell pellet. Add a volume of buffer B equal to the volume of buffer A added in step 2.
6. Mix well and repeat the centrifugation as in step 4.
7. Aspirate the supernatant and add a volume of buffer B equal to that added in step 5. Resuspend the cells by mixing thoroughly.
8. Ensure that the stabilized sample is stored at 4°C prior to use. Note: Overnight incubation of the sample in buffer B at room temperature, prior to refrigeration, may improve long-term stability.

3 Internal quality-control procedures

Numerous variables determine the outcome of flow cytometric immunophenotyping assays:

(a) specimen (collection, anticoagulant, transport, and storage);

(b) sample preparation (red cell lysis, immunostaining, any washing steps, fixation, storage until flow cytometry);

(c) immunostaining panel (choice and combination of monoclonal antibodies (mAbs) and fluorochromes, titration of mAbs);

(d) data acquisition and analysis (number of cells acquired, selection of cells of interest, discrimination between negative and positive fluorescence signals);

(e) reporting and interpretation.

The various procedures for the QC of these variables are discussed in Chapter 5. Here, we discuss the choice of useful procedure controls for the four, clinical diagnostic, immunophenotyping assays mentioned above. A procedure control is a reference sample with similar characteristics as patient samples and is processed as a patient sample. For the *quantitative assays* it makes sense to choose the reference sample such that the main cell populations under study are present in clinically relevant ranges. For example, as procedure controls, a laboratory involved in monitoring patients infected with the human immunodeficiency virus should choose samples with CD4+ T-lymphocyte counts around the level important for clinical decision-making, e.g. 200 CD4+ cells/μl. For CD34+ HPC enumeration, the target values would be around 10–20 and 200 CD34+ cells/μl, which are values that can be encountered in blood specimens during apheresis planning, in cord blood samples and in apheresis products, respectively.

3.1 When to reject a quantitative procedure control

Here, Westgard's rule (10) are of use. This strategy requires, first, that 'target values' of a given reference material are determined, e.g. by multiple (i.e. ≥20 ×) testing of that material divided over multiple (i.e. ≥5) occasions. The mean and SD of the repeated assessments are calculated, and the results of subsequent procedure controls are plotted in a Levey–Jennings chart (see *Figure 1*). According to the 'single rule', an assay is rejected as 'out-of-control' when one control measurement in the group exceeds the control limits of mean ± 2 SD. Alternatively, according to the 'multi-rule' QC procedure, an assay is rejected if:

(a) 2 consecutive control measurements exceed the same limit (i.e. >2 SD or <2 SD); *or*

(b) 4 consecutive control measurements exceed the same, more stringent, limit (i.e. >1 SD or <1 SD); *or*

(c) 10 consecutive control measurements fall on one side of the mean.

3.2 Qualitative assays

For screening the HLA-B27 antigen, the inclusion of an HLA-B27 positive sample in the assay is straightforward. The strategy for HLA-B27 negative samples is more complicated, in view of the different patterns of crossreactivity of the various anti-HLA-B27 mAbs with antigens such as HLA-B7 and B37. A safe choice

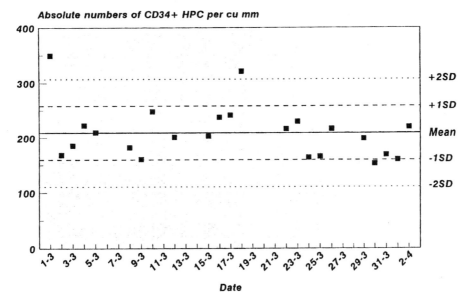

Absolute numbers of CD34+ HPC per cu mm

Figure 1 Example of a Levey–Jennings plot showing the results of absolute CD34+ HPC assessments on a stabilized apheresis product by a single laboratory. The lines indicate mean ± 1 SD and mean ± 2 SD of 20 repeated measurements. According Westgard's single rule, the 1–3 and 18–3 experiments should be rejected. According to the 'multi-rule' procedure, all runs would be acceptable.

would be both a non-crossreactive and a crossreactive HLA-B27 negative procedure control.

No adequate procedure controls have been defined for leukaemia/lymphoma immunophenotyping. The multitude of pathological immunophenotypes makes it impossible to select representative positive and negative procedure controls. Rather, pathological specimens, even when dominated by an abnormal population, will contain at least a few residual normal cells that can serve as a benchmark for the appropriate staining technique (11). For markers that are not expressed by normal cell populations, other (pathological) specimens stained in the same series may serve as controls. In addition, any fluorochrome and isotype control mAb is, by definition, not representative for the great variety of mAbs with their different protein concentrations and fluorochrome:protein ratios. Instead, any mAb panel will yield at least a few negative populations for each fluorochrome that can serve as control for non-specific mAb binding, in comparison to unstained cells as autofluorescence control.

The emerging clinical relevance of the detection of minimal residual disease in haematological malignancies (12) will require internal quality control for this, quantitative, flow cytometric assay. Suitable reference samples can be prepared by spiking malignant cells with a unique immunophenotype into a normal blood sample, followed by stabilization. The concentration should be chosen around the desired detection limit, e.g. 1 malignant cell among 10^4 leucocytes.

129

4 External quality assurance

EQAS for clinical diagnostic, flow cytometric assays have been implemented in several countries. Examples are the UK NEQAS (National External Quality Assurance Scheme) for Leucocyte Immunophenotyping (Sheffield, UK, see above), that makes use of stabilized blood specimens and accepts participants world-wide; the College of American Pathologists and FAST Systems, Inc., who operate lymphocyte immunophenotyping schemes in the USA; and SIHON-SKMI-SKZL running EQAS for leukaemia/lymphoma immunophenotyping, HLA-B27 screening, and enumeration of lymphocyte subsets and CD34+ HPC in The Netherlands and Belgium. A nationwide EQAS for lymphocyte immunophenotyping is run by a government agency in Canada. In other (European) countries, EQAS are run by the national flow cytometry (e.g. Spain, Italy) or clinical chemistry/ haematology societies (e.g. Germany).

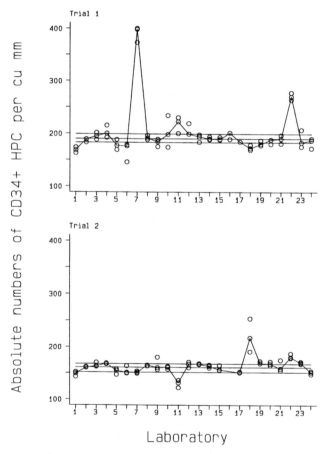

Figure 2 Triplicate assessments of absolute CD34+ HPC counts on stabilized specimens by 24 laboratories. The horizontal lines indicate the 25th, 50th, and 75th percentiles of the medians of all laboratories per trial. The median values of each of the laboratories are connected by a line to illustrate interlaboratory variation. For further discussion see text.

Participation in an EQAS is currently compulsory in some countries (e.g. USA and Canada) and voluntary in other (mostly European) countries. The implementation of laboratory accreditation will make participation in an EQAS highly desirable. This trend poses strict quality criteria to EQAS organizations. As a consequence, the EQAS themselves will become subject to accreditation procedures, such as in the UK. The materials used for send-outs should meet the same strict standards as outlined for internal controls detailed above. Importantly, an EQAS should not only have quality assessment as task, but also have an educational goal. Along the latter line, the Canadian EQAS has achieved a significant reduction in interlaboratory variation of lymphocyte subset enumeration (13), and interpretation of leukaemia/lymphoma immunophenotyping data has improved during 9 years of EQAS by SIHON (7). The increasing implementation of uniform 'single-platform' methodology for CD4+ T-cell and CD34+ HPC enumeration by the participants in the scheme of UK NEQAS for Leucocyte Immunophenotyping has also reduced the variation between laboratories (9, 14).

An example of data that can be generated in an EQAS is shown in *Figure 2*. In a trial organized by the European Working Group of Clinical Cell Analysis, 24 laboratories were requested to perform absolute CD34+ HPC counting in triplicate on two stabilized specimens. On the first specimen, laboratories 7, 11, and 22 produced outlying results, whilst laboratories 6 and 10 had poor intralaboratory reproducibility. These laboratories were provided by the organizing committee with specific recommendations to improve their technique. On the subsequent specimen, laboratories 6, 7, and 10 showed improvement, whilst laboratory 22 still produced a high outlier, although to a lesser degree than in trial 1. However, laboratory 11 still had outlying results, this time in the opposite direction, and required further assistance. Importantly, the educational efforts were reflected in an overall reduction of the interlaboratory coefficient of variation (CV) from 23% (specimen 1) to 10% (specimen 2).

5 Conclusions

The implementation and use of appropriate internal QC procedures and participation in an EQAS form the basis of an adequate quality assurance strategy. The availability and use of reference materials that:

(a) are stable over weeks to months;

(b) have interassay properties similar to those of patient specimens; and

(c) have constituents or characteristics in similar ranges as those of patient samples.

It is the responsibility of the national societies involved in EQAS organization to provide such reference materials for internal QC procedures and EQAS. Participating laboratories may use these reference materials as primary standards to calibrate their 'home-made' secondary standards for internal use.

References

1. National Committee on Clinical Laboratory Standards (1998). *Evaluation of matrix effects; proposed guideline* (document EP14-P). NCCLS, Wayne, PA.
2. Centers for Disease Control and Prevention (1997). *Morb. Mortal. Weekly Rep.*, **46,** 1.
3. National Committee on Clinical Laboratory Standards (1998). *Clinical applications of flow cytometry: quality assurance and immunophenotyping of lymphocytes; approved guideline* (document H42-A). NCCLS, Wayne, PA.
4. Gratama, J. W., Keeney, M., and Sutherland, D. R. (1999). *Curr. Protocols Cytometry*, **6,** 4.1.
5. European Federation for Immunogenetics (1996). *Standards for histocompatibility testing* (2nd version).
6. Tiwari J. L. and Terasaki, P. I. (1985). *HLA and disease association.* Springer Verlag, New York.
7. Kluin-Nelemans, J. C., Van Wering, E. R., Van 't Veer, M. B., Van der Schoot, C. E., Adriaansen, H. J., Van der Burgh, F. J., and Gratama, J. W., on behalf of the Dutch Cooperative Study Group on Immunophenotyping of Hematologic Malignancies (SIHON) (1996). *Br. J. Haematol.*, **95,** 692.
8. Barnett, D., Granger, V., Mayr, P., Storie, I., Wilson, G. A., and Reilly, J. T. (1996). *Cytometry*, **26,** 216.
9. Barnett, D., Granger, V., Storie, I., Peel, J., Pollitt, R., Smart, T., and Reilly, J. T. (1998). *Br. J. Haematol.*, **102,** 553.
10. Westgard, J. O. and Groth, T. (1983). *Am. J. Clin. Pathol.*, **80,** 49.
11. Stelzer, G. T., Marti, G., Hurley, A., McCoy, P. Jr, Lovett, E. J., and Schwartz, A. (1997). *Cytometry*, **30,** 214.
12. San Miguel, J. F., Martínez, A., Macedo, A., Vidriales, M. B., López-Berges, C., Gonzáles, M., Caballero, D., García-Marcos, M. A., Ramos, F., Fernández-Calvo, J., Calmuntia, M. J., Diaz-Mediavilla, J., and Orfao, A. (1997). *Blood*, **90,** 2465.
13. Bergeron, M., Faucher, S., Minkus, T., Lacroix, F., Ding, T., Phaneuf, S., Somorjai, R., Summers, R., Mandy, F., and the Participating Flow Cytometry Laboratories of the Canadian Clinical Trials Network for HIV/AIDS Therapies (1998). *Cytometry*, **33,** 146.
14. Barnett, D., Granger, V., Whitby, L., Storie, I., and Reilly, J. T. (1999) *Br. J. Haematol.* **106,** 1059.

Chapter 9

Measurement of cytoplasmic and nuclear antigens

Jørgen K. Larsen

Finsen Laboratory, Finsen Center, Rigshospitalet, Copenhagen University Hospital, Department 8621, Strandboulevarden 49, DK-2100 Copenhagen Ø, Denmark

1. Introduction

Flow cytometry has become the technique of choice for the quantification of a variety of intracellular molecules involved in the regulatory pathways of cellular proliferation, differentiation, and death, as well as in signal transduction and cytokine production. This development is based on recent improvements of the immunochemical methods for detecting target antigens within the cytoplasm and nucleus, supported by the increased assortment of commercially available specific antibodies and fluorochrome conjugates, new signal amplification methods, and improved instrumentation of multiparameter data acquisition and analysis.

For flow cytometric analysis of intracellular antigens it is necessary to make the cells permeable to specific antibodies, so that these and appropriate dye molecules can reach the respective antigens in the cell interior. At the same time, the antigens must be preserved in their natural, antigenic conformation, and leakage out of the cell must be prevented. It may be necessary to use fluorescence microscopy to ensure that a particular staining procedure preserves or removes the critical cellular structures and locates the stain properly. It is evident that no single protocol for cell preparation and staining is generally applicable. It must be anticipated that for nearly every new intracellular antigen or different cell type to be investigated, an appropriate staining method has to be optimized experimentally. However, there are permeabilization and fixation methods available that work for staining and analysing a variety of antigens; these can be used as first-line methods for a particular situation.

In this chapter, it has only been possible to include methods of flow cytometric quantification for a rather limited selection of intracellular antigens occurring in mammalian cells. These methods illustrate the diversity of existing methods for appropriate cell preparation and staining. It is hoped that the reader may find a starting point for optimizing a method for the particular intracellular antigen in question.

2 Methods of cell permeabilization

Permeabilization of the cell to enable diffusion of a specific antibody from the incubation medium into the cell interior is generally achieved by the use of fixatives. In the case of an antigen that is more or less soluble or has a short biological half-life, the cells must be fixed instantly. A good fixation must retain the antigen at its proper location, and must not result in destruction of the epitope to be recognized, or in aggregation of the cell suspension. Strong fixation may alter or hide an epitope, resulting in weak staining. However, the masking of an epitope can often be alleviated either by subsequent treatment with another type of fixative or with a detergent, or by heating the fixed cell suspensions in a microwave oven, or staining the cells at high ionic strength. For the detection of cytoplasmic antigens, a solid tissue must be disaggregated into a suspension of single cells prior to fixation, permeabilization, and staining (see Chapter 3). Some nuclear antigens may be stained in nuclei extracted from formaldehyde-fixed, paraffin-embedded tissue, and some may be stained in nuclear suspensions prepared with detergents from fresh or frozen tissue. The various cell-permeabilization methods presented below may be helpful in the search for practical methods for analysing new intracellular antigens or new cell types.

As first-line methods to be tested, the following four types of treatment may be recommended:

(a) formaldehyde followed by detergent;

(b) formaldehyde followed by methanol;

(c) methanol alone;

(d) detergent alone.

According to cell-permeabilization conditions that allow proper staining and analysis, the intracellular antigens are roughly distributed in the following categories (1):

(a) For antigens located close to the plasma membrane, and most so-called soluble cytoplasmic antigens:
  • use mild conditions of cell permeabilization with or without fixation. As the fixative, use formaldehyde at low concentration, temperature, and duration. Do not use acetone or alcohol.

(b) For cytoskeleton-bound and viral antigens, and some enzymes:
  • use extensive fixation with acetone, alcohol, or formaldehyde at high concentration.

(c) For antigens located in cytoplasmic organelles or granules, and nuclear antigens:
  • for this mixed group of antigens, use specific procedures to permeabilize the cell and retain or expose the antigen in a condition accessible for the antibody. These can be anything from mild to harsh procedures; examples include Ki-57 antigen and proliferating cell nuclear antigen (PCNA).

(d) For antigens that are rather indifferent to the choice of fixation/ permeabilization procedure:
  • these antigens (small in number) are accessible to antibodies after minimal treatment, but they are also stable to the harshest fixation; for example, c-myc protein.

2.1 Permeabilization by fixation

2.1.1 Alcohols and acetone

A rapid fixation is obtained by dehydration, lipid extraction, and protein precipitation. Because these fixatives dramatically alter the cell morphology, they cannot be recommended in studies of heterogeneous cell populations, such as blood and bone marrow, where light scattering defines major subpopulations. Loss of antigenicity and redistribution of antigens are often reported, but may be alleviated by fixation at low temperatures. The tendency for cell aggregation can be minimized by swirling the sample during the stepwise addition of cold fixative. Fixation with methanol, ice-cold or at $-20\,°C$, to a concentration of 50–90% in phosphate-buffered saline (PBS) is preferable to using ethanol or acetone, due to less cell aggregation and a higher precision in DNA analysis (2). For long-term storage of fixed cells, methanol is preferable to formaldehyde. Alcohols are the preferred fixatives in the combined analysis of cytoskeletal antigens and DNA (3, 4) (see *Protocol 4* and *Figure 4*). Fixation in cold methanol is used, for example, in the analysis of cyclins (5) (see *Protocol 5* and *Figure 5*), and cold acetone/methanol, for example, in the analysis of the progesterone receptor (6).

2.1.2 Aldehydes

Aldehydes fix cells by cross-linking proteins, and are good fixatives for maintaining protein structure and cell morphology. Freshly prepared formaldehyde, prepared by dissolving analytical grade paraformaldehyde (PFA) in PBS, is preferable to formalin (7). Adjustment of pH and ionic strength is important. Fixation with glutaraldehyde at high concentration induces substantial background fluorescence. Light scattering is less altered by aldehydes than by acetone and alcohols, and is reported to be unaffected by fixation in formaldehyde (4%) with a minor addition of glutaraldehyde (0.05%). Fixation with aldehydes, e.g. with formaldehyde at a concentration higher than 0.25% for longer than 10 min, results in a substantially decreased precision of DNA analysis (2). Also, cross-linking of neighbouring proteins may hinder accessibility to the antigenic site. Fixation with formaldehyde is relatively slow in the cold and partially reversible. The extent of antigenic denaturation during fixation can be controlled by adjusting the formaldehyde concentration (0.01–4%), temperature (0–20 °C), and duration of treatment (5–60 min). Most often, supplementary treatment is needed to achieve appropriate cell permeabilization for staining intracellular antigens.

2.2 Permeabilization with detergents

With detergents that dissolve membranes by forming detergent–lipid and detergent–lipid–protein mixed micelles, cells can be permeabilized without fixation.

2.2.1 Strong membrane solubilizers

The non-ionic detergents Nonidet P-40 and Triton X-100 efficiently dissolve the plasma membrane and partially dissolve the nuclear membrane at concentrations between 0.1% and 1%, even at low temperature and in physiological buffers. Therefore, these detergents are useful for the production of monodisperse suspensions of pure nuclei from most tissues. In washless methods for staining certain nuclear antigens without the use of fixatives, Nonidet P-40 or Triton X-100 is used as the single permeabilizing agent (8, 9) (see *Protocol 7*). The loss of the cell membrane and cytoplasm results in a decreased light scattering, and often also in a reduced non-specific fluorescence.

2.2.2 Mild membrane solubilizers

Lysolecithin, used under stringently controlled conditions, as well as a number of other detergents, such as Tween-20, saponin (10), digitonin, and *n*-octyl-β-D-glucopyranoside, can be used for permeabilizing the cell membrane without dissolving it completely, but giving passage for molecules of the size of IgG (approx. 150 kDa). This is valuable for staining antigens that are washed out by the more aggressive detergents Nonidet P-40 or Triton X-100, i.e. antigens located in the cytoplasm or at the cytoplasmic face of the plasma membrane, and soluble nuclear antigens.

2.3 Combined treatment with fixatives and other agents

2.3.1 Formaldehyde followed by alcohol or detergent

It is not uncommon to find that the post-treatment of formaldehyde-fixed cells with either an alcohol or a detergent improves the fluorescence staining significantly, as indicated in the following examples.

- To analyse DNA and the nuclear SV40 T-antigen, the effect of post-treatment of formaldehyde-fixed cells with methanol and/or Triton X-100 has been investigated. In comparison with methanol-fixation alone, post-treatment with Triton X-100 resulted in decreased T-antigen fluorescence due to antigen-masking, whereas post-treatment with methanol resulted in increased T-antigen fluorescence (2).

- In measurements of PCNA and the mitotic marker p105, the specific immunofluorescence of the PCNA was much higher when formaldehyde-fixation was followed by treatment with methanol, than if it was followed by treatment with Triton X-100. However, for p105 the opposite was observed. The most probable reason for this difference was thought to be that native protein structures, which hinder access for the antibody to the epitope in

PCNA, are disrupted by methanol, but not by Triton X-100, whereas structural changes caused by methanol mask the antigenic site in p105 (11).

- However, for staining many intracellular antigens, formaldehyde followed by some type of detergent is satisfactory (1). Thus, formaldehyde followed by saponin, widely used for staining intracellular cytokines, are basic components in commercial permeabilization kits (see Section 2.3.6). Mild fixation with 0.01% formaldehyde followed by digitonin was useful for intracellular staining the T-cell receptor zeta chain (12). Strong formaldehyde fixation followed by permeabilization with n-octyl-β-D-glucopyranoside enabled the staining of vimentin (13).

2.3.2 Formaldehyde mixed with other agents

A one-step treatment with a formaldehyde–detergent mixture usually results in a more extensive permeabilization of the cell membrane than can be obtained by sequential treatment with formaldehyde and the same detergent in the same concentrations. The permeability of the cell membrane is increased by using larger proportions of detergent in this mixture or a more aggressive detergent. However, these more aggressive treatments make larger gaps in the membrane before it is fixed—with an increasing risk of leakage of the interesting intracellular antigens, and also of loss of surface antigens. A formaldehyde–Tween-20 mixture has been used for the simultaneous analysis of leucocyte surface antigens and bromodeoxyuridine (BrdUrd) (14), formaldehyde/lysolecithin for c-myc oncoprotein (15), and formaldehyde/saponin for cytoplasmic immunoglobulins (16).

A periodate–lysine–formaldehyde mixture (PLP) cross-links cell carbohydrate moieties rather than proteins. Fixation with PLP is useful for simultaneously staining surface markers on leukaemia and lymphoma cells and Ki-67 (17) (see *Protocol 6*) or various oncogenes (18). PLP-fixation without and with subsequent saponin-treatment was used for measuring the surface and total (surface plus intracellular) expression, respectively, of complement receptor CR1 (19).

2.3.3 Formaldehyde followed by microwave heating

In flow cytometry as well as immunohistochemistry, heating of formaldehyde-fixed cell suspensions in a microwave oven is reported to be a rapid, simple, reproducible, sensitive, and inexpensive way of improving the staining of intracellular antigens (20, 21). Microwave heating has been successfully applied for flow cytometric measurements of cytoplasmic antigens (CD68, lipocortin-1, desmin, α-smooth actin, bcl-2, and mdr-1/gp-170) as well as nuclear antigens (PCNA, Ki-67, p53, and rb/p105) (see *Figures 1–3*). However, microwave heating works differently for cell-surface and intracellular antigens. The staining intensity of HLA class-I surface antigen was considerably decreased at microwave doses that were optimal for staining cytoplasmic CD68 and nuclear PCNA, indicating destruction of the surface antigen. Microwave heating of cell suspensions after fixation with 2% formaldehyde, as described in *Protocol 1*, resulted in measurements of the same quality as obtained with commercial permeabilization kits

(Permeafix and Fix & Perm, see Section 2.3.6) (21). Microwave heating resulted in an altered cellular light scattering, but for many types of measurements this may be compensated by the benefits of enhanced accessibility of the intracellular epitopes to specific antibodies, together with a reduced background stainability. Furthermore, microwave heating of nuclear suspensions prepared from formalin-fixed, paraffin-embedded breast tumours resulted in increased specific staining of the oestrogen and progesterone receptor (22).

Protocol 1

Antigen retrieval by microwave heating after formaldehyde-fixation[a]

Equipment and reagents

- 2% PFA freshly dissolved in PBS, pH 7.2
- Buffer: 10 mM sodium citrate buffer pH 6.0, containing 0.5% bovine serum albumin
- 50 ml propylene tubes
- Microwave oven
- 50 μm polyester mesh (e.g. Sefar, cat. no. Scrynel PE-51.HC))

Method

1. Fix the cells in suspension by pipetting samples of 10^6 cells suspended in PBS into tubes containing ice-cold, 2% PFA. Vortex immediately.

2. After 15–30 min of fixation, centrifuge at 500 g for 5 min at room temperature and discard the supernatant. Resuspend the cells in 10 ml of buffer in an unsealed 50 ml propylene tube.

3. Heat the cell suspension in a microwave oven at the maximum power setting (600–800 W) until the temperature is close to the boiling point[b]. Immediately chill the samples on ice for 10 min.

4. Centrifuge the cells (500 g, 5 min, 4° C), discard the supernatant, resuspend the cells in PBS and filter through a 50 μm mesh before staining the intracellular antigens with the respective antibodies.

[a] Modified from the methods given in ref. 20 and 21.

[b] Find out in advance how long you need to heat the appropriate number of 10 ml samples to boiling point. In pilot experiments for staining the intracellular antigen in question, try various microwave heating times in the range 0.1 to 1 × the microwave heating time required for boiling, and then choose the best setting for further measurements.

2.3.4 Detergent followed by fixative

The process of cell permeabilization with mild detergents may be arrested by fixation, e.g. for the measurement of intracellular interleukin-2 the permeabilization with lysolecithin was arrested by fixation with formaldehyde (23). If soluble PCNA is eliminated by cell lysis with Triton X-100 or Nonidet P-40 prior

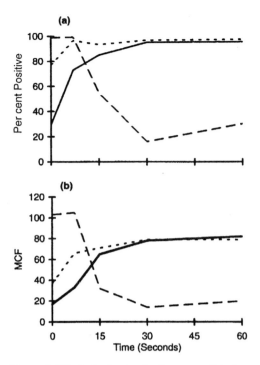

Figure 1 Effect of microwaving time on cell-surface, cytoplasmic, and nuclear-antigen labelling. U937 cells were fixed in 2% PFA and then microwaved for 0, 7, 15, 30, and 60 sec. The monoclonal antibody labelling of cell-surface (MHC class I, – – –), cytoplasmic (CD68, ·····) and nuclear (PCNA, —) antigens was measured by flow cytometry. The results from three experiments are expressed as: (a) percentage positive and (b) mean channel fluorescence (MCF). (Reproduced from ref. 20, Figure 1, with permission from the publisher and the authors.)

to fixation with cold methanol, the S-phase specific, chromatin-bound PCNA can be stained with the PC-10 antibody, see Section 4.9.3.

2.3.5 Nuclear extraction from archival material

Retrospective analysis may be possible on archival, formaldehyde-fixed and paraffin-embedded tissue. Proteases such as pepsin, trypsin, or subtilisin are used to digest re-hydrogenated, thick (30–100 μm) sections of paraffin-embedded material in order to produce nuclear suspensions. This technique allows simultaneous staining of DNA and a series of nuclear antigens, e.g. p53 (24), p105, and c-myc protein. A combination of the procedures for enzymatic digestion and microwave antigen retrieval was successful in measuring oestrogen and progesterone receptors together with DNA (22), and may be advantageous in detecting other nuclear antigens as well.

2.3.6 Commercially available permeabilization kits

Fixation–permeabilization kits, meant for the routine measurements of intracellular differentiation antigens and cytokines, are now available from several

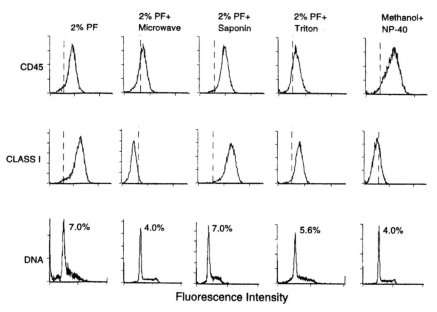

Figure 2 Comparison of different methods in the detection of cell-surface antigens and DNA content analysis by flow cytometry. U937 cells were fixed in 2% paraformaldehyde (PF) and then followed by permeabilization using microwaving, saponin, or Triton X-100 treatment, or were fixed in methanol followed by permeabilization using detergent Nonidet P-40. Microwave treatment, like methanol, produced an optimal DNA staining as shown by CV % of G_0/G_1 peak, while cell-surface antigens (CD45 and class I) were reduced. The broken line represents the positive cut-off point as determined by the negative-control antibody. (Reproduced from reference 20, Figure 2, with permission from the publisher and the authors.)

companies (see *Table 1*). The detailed chemical composition is proprietary, but most of these kits probably contain formaldehyde for fixation and mild detergents such as saponin for permeabilization. Considering their potential applications in routine haematology and immunology (see Section 4.2), it has been important to set up comparative testing of commercial and custom-defined protocols for fixation and permeabilization (25–27). Recently, under the auspices of the European Working Group on Clinical Cell Analysis, a panel of six commercially available permeabilization kits (Cytofix/Cytoperm, Fix & Perm, Intraprep, Intrastain, PermaCyte, and Permeafix) were evaluated for three important cytoplasmic leucocyte antigens: myeloperoxidase, cyCD3, and cyCD79a (28). For each of the antigens, the use of Permeafix resulted in the largest specific immunofluorescence intensity. Intraprep, Intrastain, and Fix & Perm qualified with different rankings according to the antigen in question. Cytofix/Cytoperm induced a relatively high autofluorescence, resulting in a lower signal-to-noise ratio than with most of the other kits. PermaCyte was not included in the detailed study, because changes in the light-scatter parameters made it impossible to clearly distinguish the different major leucocyte populations.

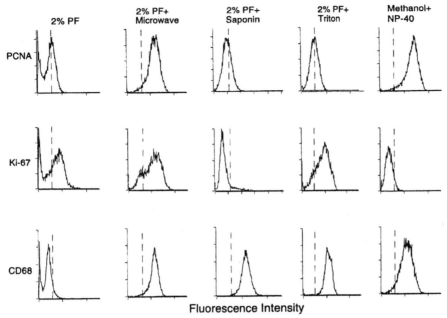

Figure 3 Comparison of different methods in the detection of cytoplasmic (CD68) and nuclear antigens (PCNA and Ki-67) in U937 cells by flow cytometry. U937 cells were fixed in 2% paraformaldehyde (PF) and then followed by permeabilization using microwaving, saponin or Triton X-100 treatment, or were fixed in methanol followed by permeabilization using detergent Nonidet P-40. Microwave treatment produced the best results for both cytoplasmic and nuclear antigen detection compared to the other methods used. The broken line represents the positive cut-off point as determined by the negative-control antibody. (Reproduced from reference 20, Figure 3, with permission from the publisher and the authors.)

Table 1 Commercially available cell-fixation/permeabilization reagents

Name of reagent	Producer	Procedure	References
Cytofix/Cytoperm	Pharmingen	1 step	28
FACS Lysing and Permeabilising Solutions	Becton Dickinson	1–2 steps	25–27, 36
Fix & Perm	Caltag Laboratories	2 steps	21, 25–28
Intraprep	Beckman Coulter	2 steps	26, 28
Intrastain	DAKO	2 steps	28
Optilyse B Lysing Solution	Immunotech	1 step	27
PermaCyte	Bio-Ergonomics	3 steps	28
Permeafix	Ortho Diagnostic Systems	1 step	21, 25, 27, 28, 79

2.4 Permeabilization without fixatives or detergents

For flow cytometric measurements on live cells, it is difficult to achieve a quantitative staining of intracellular antigens, because antibodies have to be transferred to the cell interior during a temporary permeabilization of the cell membrane in order to avoid disturbance of cell functions. A stoichiometric staining of the entire cell population can rarely be obtained, due to the variability in permeabilization efficiency for the individual cells and the relatively short period of permeabilization possible. A considerable fraction of the cells expressing the antigen in question may not be stained at all, and may be falsely classified as antigen-negative cells. In addition, a surplus of free antibody may be trapped inside the cell, falsely indicating a positive staining. However, the techniques for flow cytometric analysis of intracellular antigens in viable cells are of potential interest and are being rapidly developed.

2.4.1 Liposomes

The intracellular introduction of antibodies by fusing antibody-loaded liposomes with cells is a promising new technique in flow cytometry. The application of uniformly sized, antibody-conjugated liposomes loaded with large amounts of fluorescence-conjugated antibody and small magnetic particles enables flow cytometric detection of weakly expressed intracellular antigens such as cytokines, due to a 100–1000-fold increased staining as compared to conventional techniques. These reagents also enable magnetic sorting of these cell subpopulations (29, 30); a technique currently being further developed by the Miltenyi Biotec company.

2.4.2 Other techniques

Electroporation of cells, a technique widely used for gene transfection, has also been used in flow cytometry, see Chapter 15. Moderate treatment with proteases such as collagenase, pronase, or trypsin, under similar conditions as used for cell detachment and replating in cell culture, makes cells hyperpermeable to the entry of various proteins (ranging in size from insulin to thyroglobulin), without essential loss of viability, but with the risk of surface-antigen loss (31). The pore-forming protein, streptolysin-O, might be useful, in line with the effects of mild detergents (32).

3 Methods of cell staining and analysis

For the optimal immunochemical staining of intracellular antigens, it is recommended that the amount of antibody and the number of cells per sample is standardized by titration experiments (7). It may be difficult to reach a saturation level in fluorescence intensity, because an increase in the dose of antibody may result in a relatively higher increase in the contribution of non-specific fluorescence from irrelevant cellular components. Often, the non-specific staining may be reduced by adding a blocking agent like bovine serum albumin (BSA). In contrast to the conditions for staining of surface antigens, the use of very large molecular constructs for the identification and visualization of intracellular antigens may be difficult, due to steric hindrances for access to the specific epitopes. Thus, staining may be possible with an antibody conjugated with FITC (0.4 kDa), but not with R-phycoerythrin (PE) (240 kDa) or with tandem-molecules containing PE or allophycocyanin (100 kDa). Likewise, the use of IgG type antibodies (150 kDa) are preferable to the use of IgM type (950 kDa). However, such difficulties may be alleviated by choosing another procedure for fixation and permeabilization.

The specificity of the antibody applied must primarily be controlled by running additional samples stained with an irrelevant antibody of the same isotype (see Chapter 5). However, the entire set-up of the staining and measurement procedures should also be validated by analysing simultaneously stained control samples with known proportions of antigen-positive and -negative cells. Such experimental or biological controls may be more informative than isotype controls, when 'positively stained' cells are clearly separated as a cluster from a 'negatively staining' subpopulation in the same sample. However, isotype controls may indicate a non-specific staining of the 'negatively staining' cells. Furthermore, the use of isotype controls is essential when positive and negative subpopulations are not readily identified, because the distribution is continuous and just shifting a little in the direction of dimmer or brighter fluorescence. Isotype controls may disclose a different level of non-specific background fluorescence in the individual subpopulations of a heterogeneous sample, and this level may change according to cell activation or degradation (33). Fluorescent standard microspheres or the chicken and trout erythrocytes, which are often added as internal DNA references, may also be useful as scale markers for the antigen staining.

3.1 Simultaneous measurement of intracellular and surface antigens

Although some structures remain unaltered after fixation, it is unwise to assume that surface antigens, which are mostly defined on viable cells, will survive intact for specific staining after fixation. For the combined study of cell-surface and intracellular antigens, it is therefore common practice to stain the surface antigens first (see Chapter 5), then fix and permeabilize, and finally to stain the intracellular antigens, e.g. leucocyte differentiation antigens (34), cytokines (35, 36) (see also Chapter 7, Section 10), vimentin (13), terminal deoxynucleotidyl transferase (Tdt) (37), or Ki-67 (17, 38) (see *Protocol 6*).

3.2 Simultaneous measurement of intracellular antigens and DNA

Simultaneous measurement of the DNA content enables analysis of the differential expression of intracellular antigens in relation to the cell-cycle variations and heterogeneity in DNA ploidy, e.g. for cytokeratins (3, 4, 39) (see *Protocol 4* and *Figure 4*), cyclins (5, 40, 41) (see *Protocol 5* and *Figure 5*), Ki-67 (9, 10) (see *Protocol 7*), PCNA (8, 42, 43) (see *Protocol 8* and *Figure 6*), c-myc oncoprotein (15), p53 (24, 44), and hormone receptors (6, 45), see also Sections 4.7–4.11.

For bivariate fluorescence analysis based on laser excitation at 488 nm, DNA is stained with propidium iodide (PI) and the intracellular antigen with an FITC-conjugated monoclonal antibody, or indirectly with an mAb followed by a secondary polyclonal, FITC-conjugated antibody (7, 11) (see *Protocols 4, 5, 7,* and *8*).

For the triple-analysis of two antigens plus DNA, antibodies conjugated with FITC and PE may be used together with either PI (46), 7-aminoactinomycin D (7-AAD) (47), or a mixture of PI and TO-PRO-3 for DNA staining (48) (see *Protocol 2*). However, correct compensation for the considerable spectral emission overlap between PE and PI is rather difficult if the PE-stained antigen is only weakly expressed. The emission overlap is substantially less between PE and 7-AAD, but staining with 7-AAD is more dependent on variations in nuclear chromatin structure and is therefore less DNA stoichiometric than staining with PI. TO-PRO-3 cannot itself be excited at 488 nm. However, as both the PI and TO-PRO-3 molecules contain the propidium component, they intercalate between the DNA base pairs in positions so close to each other that energy transfer occurs between them. Using a mixture of PI and TO-PRO-3 for DNA staining, it is therefore possible to preserve the good stoichiometry of the propidium–DNA interaction and at the same time reduce the problems of spectral emission overlap, because the majority of the DNA fluorescence emission is then shifted to a higher wavelength. Using flow cytometers equipped with additional light sources, it is possible to extend the number of antigens that can be measured simultaneously with DNA further, by separate excitation of the DNA-dyes DAPI, Hoechst 33342, or TO-PRO-3.

Protocol 2

Staining two intracellular antigens and DNA, using PI and TO-PRO-3[a]

Equipment and Reagents

- 1% PFA in PBS, pH 7.2
- Two monoclonal antibodies of different Ig subclasses (e.g. 2.5 µg/ml anti-p53 antibody, DAKO DO-7, IgG2b, and 5.0 µg/ml anti-cytokeratin 8/18, Becton Dickinson, CAM5.2, IgG2a)
- Negatively staining isotype antibodies for control
- PBS/BSA: 0.5% BSA in PBS
- Appropriate FITC- and PE-labelled secondary antibodies (goat–anti-mouse subclass-specific polyclonal antibodies)
- DNA-staining solution: 100 µM PI (Sigma or Molecular Probes), 2.0 µM TO-PRO-3 (Molecular Probes), and 0.1% RNase (Sigma R-4875) in PBS/BSA
- Sample tubes as required for flow cytometer

Method

1. Wash about 10^6 tumour cells (harvested as described in Chapter 3) twice with cold PBS (centrifuge at 300g for 5 min, 4°C) and subsequently fix with freshly prepared 1.0% PFA in PBS, added dropwise with constant swirling. Fix the cells for 5 min on ice, then add 1.0 ml PBS/BSA.

2. Centrifuge, (300 g, 5 min, 4°C) resuspend the cell pellet by vortexing, and permeabilize by adding 1.0 ml of cold 100% methanol (−20°C) drop-wise while swirling. Store cells in the freezer at −20°C for 10 min, centrifuge, and wash twice with PBS/BSA.

3. To one sample, add 100 µl of the appropriate monoclonal antibodies. To another sample, add negatively staining isotype antibodies for control.

4. Incubate for 30 min at 4°C. Wash twice with 1.0 ml cold PBS/BSA.

5. Add 100 µl of the appropriate FITC- and PE-labelled secondary antibodies diluted 1:200 in PBS/BSA, and incubate for 30 min at 4°C. Wash once with PBS/BSA.

6. Add 0.5 ml of the DNA-staining solution, and, to activate the RNase, incubate the cells at 30 min at 37°C, prior to flow cytometry using 488 nm excitation.

[a] Modified from the method given in ref. 48.

3.3 Simultaneous measurement of intracellular antigens and bromodeoxyuridine

The immunochemical detection of BrdUrd incorporated by DNA-synthesizing cells, as described in Chapter 10, can be combined with the measurement of intracellular and cell-surface antigens. As a dynamic measure of cell-cycle progression, BrdUrd incorporation is important in studies of cytokine-related cell activation, eventually leading to cell proliferation and/or differentiation. In addition to fixation and permeabilization, BrdUrd immunodetection requires partial denaturation of the DNA to the single-stranded state to allow the anti-BrdUrd antibody to bind the incorporated thymidine analogue into DNA. For the combined measurement of BrdUrd incorporation and cytokine expression in T-cell subsets from activated peripheral blood mononuclear cells, DNA was partially denatured using DNase I (49) (see *Protocol 3*). For the combined measurement of BrdUrd incorporation, DNA content, and cyclin B1 expression in human cancer cell lines, DNA was denatured by treatment with 2–3 M HCl (50). Another DNA denaturation method that might be useful for the simultaneous analysis of BrdUrd and intracellular antigens is the so-called SBIP (strand breaks induced by photolysis) method, based on selective DNA strand-break induction by ultraviolet photolysis at sites that contain incorporated BrdUrd, followed by Tdt-facilitated DNA strand-break labelling with bromodeoxyuridine triphosphate (BrdUTP), and immunostaining of the BrdUTP with anti-BrdUrd antibody (51).

Protocol 3

Staining of BrdUrd, CD4, and intracellular cytokine[a]

Reagents

- Staphylococcal enterotoxin
- Brefeldin A (Sigma)
- BrdUrd (5-bromo-2'-deoxyuridine, Sigma)
- FACS Permeabilizing Solution (Becton Dickinson)
- FACS buffer: 0.5% BSA and 0.1% sodium azide in PBS

Protocol 3 continued

- DNase I (Sigma, cat. no. DN-25 400–600 Kunitz units/mg)
- 1% PFA in PBS
- PE-conjugated antibody (e.g. against CD4, CD69, CD25, interferon-γ, or IL-2) (Becton Dickinson)
- FITC-conjugated anti-BrdUrd antibody (Becton Dickinson or Harlan Sera Lab)
- PerCP-conjugated anti-CD4 antibody (Becton Dickinson)
- Negatively staining isotype antibodies for control

Method

1. Activate peripheral blood mononuclear cells (e.g. with staphylococcal enterotoxin), and for the last 6 h of the culture period add 10 μg/ml brefeldin A and 60 μM BrdUrd (protect from strong light exposure to avoid BrdUrd decomposition).

2. Centrifuge (500 g, 5 min, 4°C) the harvested cells. Resuspend for fixation and permeabilization with 0.5 ml of FACS Permeabilizing Solution for 18 h at 4°C. Wash the permeabilized cells twice with FACS buffer.[b]

3. Incubate 100 μl aliquots of 2–4 × 10⁵ permeabilized cells for 1 h at room temperature with 20 μl of a mixture containing 10 mg/ml DNase I, FITC-conjugated anti-BrdUrd antibody, the desired PE-conjugated anti-cytokine antibody, and PerCP-conjugated anti-CD4 antibody. To another sample, add corresponding negatively staining isotype antibodies for control.

4. Wash with PBS, and fix the samples with 300 μl of 1% PFA in PBS prior to triple-fluorescence cytometric analysis.

[a] Modified from the method given in ref. 49.

[b] Alternatively, add 1 ml of PBS, containing 1% PFA and 0.01% Tween-20, resuspend carefully, and store overnight at room temperature, then wash twice with PBS (containing calcium and magnesium) before staining (14).

4 Examples of intracellular antigens that can be measured by flow cytometry

4.1 Plasma-membrane associated antigens

The cytoplasmic face of various receptors or second messengers can be studied by flow cytometry, as important molecules in the interaction between extracellular, activating factors and endogenous factors in signal transduction pathways. The quantity of intracellular complement C3b/C4b receptors in human neutrophil granulocytes has been estimated by the increment in immunofluorescence between PLP-fixed cells before and after saponin-permeabilization (19). Some multidrug-resistance associated membrane proteins are identified with antibodies against epitopes on the cytoplasmic face of the molecule (52). The intra- and extracellular exposure of domains of the beta-amyloid precursor protein (βAPP), related to Alzheimer's disease, has been investigated (53).

4.2 Leucocyte differentiation antigens

In immunophenotyping for haematological diseases, flow cytometric characterization of the leucocyte differentiation lineages and the functional maturation
of cells has become of major importance, and is based on the detection of
intracellular as well as surface antigens (see also Chapter 5). A series of commercial kits for intracellular staining are available (see Section 2.3.6). The
earliest and highly specific markers for the different lymphoid and myeloid
haematopoietic cell lineages are frequently absent from the cell surface, but are
usually detectable at the intracellular level. Multiparameter flow cytometric
analysis including such intracellular markers has been of great value for the
establishment of cell lineage in human acute leukaemia, for the classification of
morphologically undifferentiated leukaemias, and for the identification of a
subset of leukaemias with blast cells displaying markers characteristic of more
than one haematopoietic cell lineage. In addition, these markers have been
useful for detecting residual disease. Among these cytoplasmic markers,
particularly myeloperoxidase, cyCD3 and cyCD79a have become important (28).

4.3 Cytokines

The flow cytometric analysis of the expression of various cytokines and
chemokines in leucocyte subsets has recently become an important tool in
haematological and immunological research (35, 36, 54). One application is the
analysis of T-cell immunoreactivity into a restricted (T_{H1}- or T_{H2}-like cells) or
unrestricted (T_{H0}-like cells) pattern, according to the types of cytokines (IL-4, IL-
10, interferon-γ, etc.) that are expressed after experimental stimulation of the
cells and blocking of the secretion via the Golgi apparatus. For immunochemical
staining of intracellular cytokines and chemokines, the permeabilization protocols are similar to those used for intracellular leucocyte differentiation
antigens (see Section 4.2), mainly based on fixation with formaldehyde and
permeabilization with mild detergents such as saponin (see also Sections 2.3.6,
2.4.1, 3.1 and 3.3, *Protocol 3*, as well as Chapter 7, Section 10).

4.4 Enzymes

An enzyme may be quantified with regard to its activity by measurement of the
conversion of a non-fluorescent substrate to a fluorescent reaction product (see
Chapter 2)—and also, with regard to the cellular content of enzyme molecules,
by immunochemical measurement using specific antibodies. Immunochemically, and based on a variety of methods for cell fixation and permeabilization, a
large number of enzymes can be quantified (1), e.g. a series of lysosomal and
mitochondrial enzymes, including metabolic enzymes as well as caspases, which
are involved in apoptotic cell death (see Chapter 14). Inducible nitric oxide
synthase has been measured in formaldehyde- and saponin-treated adjuvant
arthritis synovial macrophages (55). Several of the many enzymes involved in
DNA synthesis can be measured, e.g. DNA polymerase-α (56), ribonucleotide
reductase (57), Tdt (37), DNA-methyltransferase (58), and DNA topoisomerase-II
(59), the latter being important in investigations of drug resistance in cancer
chemotherapy.

4.5 Hormones and receptors

Hormones and their receptors can be measured, e.g. growth hormone, luteinizing hormone, and prolactin (1), as well as the receptors of glucocorticoid hormone (60) and of oestrogen and progesterone (6, 45, 61). Intracellular receptors can also be measured using fluorochrome-conjugated ligands, e.g. D-mannose-specific receptor (62).

4.6 Fetal haemoglobin

The detection of fetal cells in maternal blood represents an important area of laboratory support to the obstetrical management of women. In addition, the isolation of fetal cells has developed as a means for the non-invasive detection of fetal genetic abnormalities. Following various enrichment procedures, flow cytometric detection of fetal cells, according to staining with antibodies against fetal haemoglobin, is important for estimating the proportion of fetal cells in maternal blood (63).

4.7 Cytoskeleton

Cytoskeletal proteins can easily be immunochemically stained in fixed cells and detected by flow cytometry with relatively strong fluorescence signals (1, 3). Some of the cytoskeletal proteins are diagnostic for specific differentiation lineages: cytokeratins are detected in epithelial cells, and vimentin (13) in cells of mesenchymal origin. Others—like actin, tubulin (64), kinetochore protein, and intermediate filaments—are more generally distributed. Keratin, a specific set of cytokeratins, as well as other antigens are recognized in epidermal cells (3). The flow cytometric detection of individual, DNA-aneuploid, carcinoma-cell subpopulations in tumours and estimation of their S-phase fraction can be significantly improved when based on the simultaneous staining of DNA and cytoplasmic cytokeratins (4) (see *Protocol 4* and *Figure 4*). The critical procedure in

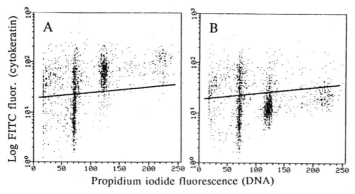

Figure 4 Analysis of cytokeratin and DNA content in cells of human mammary adenocarcinoma MDA 231, growing subcutaneously in a nude mouse (see *Protocol 4*). Dot plots represent 10^4 cells, stained with: (A)–anti-cytokeratin antibody (DAKO MNF-116), (B) isotype control antibody. Gates on light-scatter were used to exclude debris and aggregates. (Reproduced with permission from Jacob Larsen.)

this type of analysis is the disaggregation of tumour tissue into a single-cell suspension. Cytokeratins may even be recognized in cells extracted from paraffin-embedded tissue (65).

Protocol 4

Staining of cytokeratin and DNA in tumour cells

Equipment and reagents

- 50 μm polyester mesh (e.g. Sefor, cat. no. Scrynel PE-51-HC)
- Anti-cytokeratin antibody (DAKO, cat. no. MNF-116, or Becton Dickinson, cat. no. CAM5.2) and isotype control antibody
- FITC-conjugated rabbit–anti-mouse antibody (DAKO, cat. no. F-313)

- PBS/NP-40/BSA: 0.1% Nonidet P-40 (BDH) and 0.1% BSA in PBS
- DNA-staining solution: 50 μg/ml PI (Sigma or Molecular Probes), 0.2 mg/ml RNase (Sigma, cat. no. R-4875), and 0.1% Nonidet P-40, in PBS

Method

1. Prepare a suspension of single cells from the tumour biopsies, using enzymatic and/or mechanical methods for tissue disassociation (see Chapter 3). Wash the cells in PBS.

2. Fix the cells in suspension by adding 2 ml of ice-cold 70% ethanol to samples of $0.5-1 \times 10^6$ cells in 200 μl of PBS while swirling. Store the fixed cells in the refrigerator for at least 30 min. Wash in PBS before staining. Filter the suspension through 50 μm polyester mesh.

3. Add 50 μl-anti-cytokeratin antibody diluted 1:10 in PBS/NP-40/BSA. To another sample, add an equivalent amount of the isotype control antibody.

4. Incubate for 30 min at room temperature. Wash once in PBS/NP-40/BSA.

5. Add 50 μl of the FITC-conjugated rabbit–anti-mouse antibody diluted 1:10 in PBS/NP-40/BSA and incubate at room temperature for 30 min. Wash once in PBS/NP-40/BSA.

6. Add 400 μl DNA-staining solution and leave for at least 30 min prior to flow cytometry.

4.8 Cyclins

The progression of cells through the respective phases and checkpoints of the cell cycle is controlled by the cyclin-dependent protein kinases (CDKs) and their regulatory partners, the cyclins. In flow cytometry, the scheduled expression of cyclins together with DNA content are important markers for mapping cell-cycle events (66, 67) (see *Protocol 5*). Normally, increased expression of type-D cyclins is seen at the G_0-G_1 restriction point, cyclin E is increased in early S phase, and cyclin B1 in G_2 phase and mitosis (see *Figure 5*). However, the cyclin-D

JØRGEN K. LARSEN

subtypes D1, D2, and D3 seem to be specific for different tissues, and in tumour cells the cyclin expression may be unscheduled (5, 41, 68).

Protocol 5

Staining of cyclin and DNA[a]

Reagents

- PBS/BSA: 1% BSA in PBS
- PBS/Triton: 0.25% Triton X-100 in PBS
- Propidium iodide (Sigma or Molecular Probes)
- RNase A (Sigma)
- FITC-conjugated goat–anti-mouse IgG antibody (Sigma)
- Anti-cyclin antibodies
- 1% PFA in PBS

A. Cyclins A, B1, or E

1. Fix cells in suspension by pipetting samples of $1-2 \times 10^6$ cells suspended in 1 ml PBS into tubes containing 10 ml of 100% methanol or 80% ethanol at $-20\,°C$. Store fixed cells at $-20\,°C$ for 2-24 h.

2. Centrifuge (500 g, 5 min, 4°C) the cells, rinse once with PBS/BSA, centrifuge again, and suspend the cell pellet in 1 ml of PBS/Triton. Keep on ice for 5 min, then add 5 ml of PBS and centrifuge.

3. Suspend the cell pellet in 100 µl of PBS/BSA, containing the respective cyclin antibody[a] at a concentration of 0.25 µg/10^6 cells (or for control samples an equivalent amount of isotype control antibody). Incubate at 4°C overnight.

4. Rinse the cells with PBS/BSA, centrifuge, and suspend the cell pellet in 100 µl of PBS/BSA containing FITC-conjugated goat–anti-mouse IgG antibody (diluted 1:40). Incubate for 30 min at room temperature.

5. Rinse the cells with PBS/BSA and suspend the cell pellet in a solution containing 10 µg/ml propidium iodide (PI) and 1 mg/ml of RNase A in PBS. Incubate for 20 min at room temperature before measurement.

B. Cyclins D1, D2, or D3

1. Fix the cells in suspension by pipetting samples of $1-2 \times 10^6$ cells suspended in 1 ml PBS into tubes containing 10 ml of ice-cold 1% PFA in PBS, for 15 min.

2. Rinse the cells with PBS, centrifuge, and post-fix by suspending the cell pellet in 5 ml of 80% ethanol at $-20\,°C$. Store fixed cells at $-20\,°C$ for 2-24 h.

3. Proceed as in Part A, steps 2-5.

[a] Modified from the methods given in refs 5 and 41, where certain–anti-cyclin antibodies are recommended.

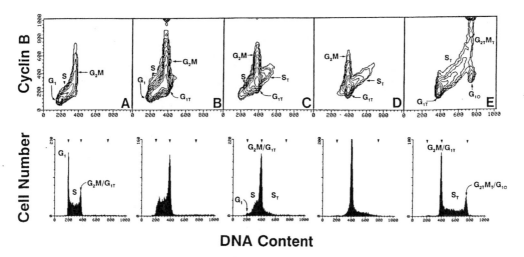

DNA Content

Figure 5 Changes in the bivariate distribution of DNA content/cyclin B expression of MOLT-4 cells treated with 0.1 μM staurosporine for 4 (B), 6 (C), 8 (D), and 12 h (E); (A) untreated culture. Top, appearance of cyclin B-negative cell population with G_2+M DNA content (tetraploid; G_{1T}) is already evident after 4 h, and cyclin B-negative cells with octaploid DNA content (G_{10}), after 12 h. Bottom, DNA-content frequency histograms of the same cultures. Arrows, the position of cells with a 2C, 4C, and 8C DNA content, respectively. (Reproduced from reference 77, Figure 3, with permission from the publisher and the authors.)

4.9 Other proliferation-associated antigens

4.9.1 Mitotic cell antigens

A cell population's birth rate can be estimated by experimental arrest of cells in mitotic metaphase, using spindle poisons such as colcemid or vinblastine. Mitotic cells can be discriminated immunochemically from interphase cells by their relatively higher expression of the nuclear antigens p105 (8, 67) and MPM-2 (8), or the cytoplasmic antigen AF-2 (69).

4.9.2 Ki-67

In normal tissues the Ki-67 antibody identifies a nuclear antigen in cycling cells, as opposed to non-cycling cells (70). Several staining procedures have been described for flow cytometric, dual-parameter analysis of Ki-67 expression vs. surface antigens (17, 38) (see *Protocol 6*), DNA content (9, 10), or PCNA (8, 47). Double-staining of Ki-67 and DNA in nuclear suspensions of detergent-treated, unfixed cells is simple and allows the measurement of DNA with high precision (8, 9) (see *Protocol 7*). However, when using monoclonal Ki-67 antibody (DAKO) on detergent-treated, unfixed cells of non-haematopoietic origin, adverse Ki-67 staining in the cytoplasm of non-cycling cells may become a disturbing factor. In flow cytometric measurements on mammary carcinoma cells (71), all tested Ki-67 type antibodies (monoclonal Ki-67, MIB-1, and Ki-S5, and polyclonal Ki-67) worked equally well on methanol-fixed MCF-7 cells, but only the MIB-1 and Ki-S5 epitopes were retained in cell suspensions mechanically derived from fresh-frozen tissue. Furthermore, only the antibody Ki-S5 showed specific nucleolar staining patterns in cell suspensions prepared by trypsin digestion of formalin-fixed, paraffin-embedded tissue sections.

Protocol 6

Staining of Ki-67 and cell-surface antigen[a]

Reagents

- PLP: 10 mM sodium periodate, 75 mM lysine, and 2% PFA in 37 mM PBS at pH 7.4
- FITC-conjugated Ki-67 antibody (DAKO, cat. no. F-788)
- PBS/BSA: 1% BSA in PBS

- FITC-conjugated isotype control antibody (DAKO, cat. no. X-927)
- PE-conjugated monoclonal antibody against appropriate surface antigen
- PE-conjugated control antibody

Method

1. Prepare samples of 10^6 cells suspended in 100 μl of ice-cold PBS/BSA.

2. Add 50 μl of PE-conjugated surface-marker antibody diluted in PBS/BSA, or the equivalent amount of isotype control antibody.

3. Incubate for 30 min at 4°C. Wash twice in PBS/BSA.

4. Fix the cells in suspension by adding, for 15 min at −10°C, 1 ml of PLP (pre-cooled to −10°C). Wash twice with PBS.

5. Add 50 μl of FITC-conjugated Ki-67 antibody diluted 5–10-fold in PBS/BSA, or the equivalent amount of FITC-conjugated isotype control antibody.

6. Incubate for 30 min at 4°C. Wash twice with PBS before flow cytometric analysis.

[a] Modified from the method given in ref. 17.

Protocol 7

Washless staining of Ki-67 and DNA[a]

Reagents

- Freezing buffer: 250 mM sucrose, 5% (v/v) dimethylsulfoxide, 40 mM Na citrate, pH 7.6
- Lysis–DNA staining solution: PBS plus 0.5% (v/v) Nonidet P-40 (BDH), 20 μg/ml PI (Sigma or Molecular Probes), 0.2 mg/ml RNase (Sigma, cat. no. R-4875), 0.5 mM EDTA, at pH 7.2

- Monoclonal Ki-67 antibody (DAKO, cat. no. M-722)
- Isotype control antibody (DAKO, cat. no. X-931)
- FITC-conjugated rabbit–anti-mouse antibody (DAKO, cat. no. F-313) diluted 10-fold in PBS with 5% normal rabbit serum (DAKO, cat. no. X-902)

Method

1. Centrifuge samples of approx. 10^5 cells in PBS. Add 50 μl of ice-cold freezing buffer to the cell pellet and resuspend. Freeze the samples for storage at −80°C.

Protocol 7 continued

2. Thaw the sample. Stain the cells by stepwise addition of reagents without any intermediate or final washing or filtration, resulting in a final volume of 250–300 μl stained sample ready for flow cytometry. Keep the sample tubes in an ice-bath.

3. Add 200 μl of the lysis–DNA staining solution and incubate on ice for 15 min.

4. Add 25 μl of monoclonal Ki-67 antibody diluted 1:10 in PBS/BSA, or the equivalent amount of isotype control antibody. Incubate on ice for at least 15 min.

5. Add 25 μl of FITC-conjugated rabbit–anti-mouse antibody. Incubate on ice for at least 15 min prior to flow cytometry.

[a] Modified from the method given in ref. 72.

4.9.3 PCNA

PCNA functions as a part of the DNA polymerase holoenzyme at the DNA replication forks for semi-discontinuous DNA synthesis. The current model for the function of PCNA suggests that PCNA acts as a sliding clamp that allows DNA polymerases (pol-δ/ε) to move rapidly along the DNA while remaining topologically bound to it. PCNA has also been shown to participate in nucleotide-excision DNA repair. In proliferating cells, PCNA cycles between a chromatin-bound, detergent-insoluble state in S phase, and a diffuse soluble state when DNA is not being replicated or repaired (73). The solubility is associated with phosphorylation of PCNA, and the loading of PCNA on to DNA seems sufficient to explain the detergent insolubility and association with replication sites. PCNA has initially been identified with a human auto-antiserum, but several monoclonal antibodies are now available, e.g. 19A2, 19F4, and PC-10 (42, 74). However, according to the choice of antibody and of the procedure for fixation and staining, monoclonal–anti-PCNA antibodies may identify either the cells in S phase only, or the entire cycling, or perhaps even part of the non-cycling, cell population (see *Figure 6*). In unfixed nuclei, some PCNA auto-immune antisera specifically identify S-phase cells (8). With the monoclonal PCNA antibody PC-10, exclusively S-phase cells are stained if the cells are treated with detergent prior to fixation with methanol (see *Protocol 8* and *Figure 6C* and *F*), whereas all cycling cells are stained if the cells are directly fixed (42, 43) (see *Figure 6A* and *D*). Combined analysis of Ki-67 and S-phase associated PCNA allows the discrimination of G_0, G_1, S, G_2, and M cells without the simultaneous staining of DNA (8).

4.10 Cell death-associated antigens

An increasing number of antibodies against the various antigens involved in apoptotic cell death, e.g. caspases and the bcl-2 family, are now available for flow cytometric investigations, see also Chapter 14. Some stress-induced proteins, like the heat-shock protein 70 (hsp70), can be quantified by flow cytometry (75).

Protocol 8

Staining of PCNA and DNA[a]

Reagents

- Lysing buffer: 0.5% Triton X-100, 0.2 µg/ml EDTA, and 1% BSA in PBS
- Monoclonal PC-10 antibody (DAKO, cat. no. F-863)
- Isotype control antibody (DAKO, cat. no. X-933)
- DNA-staining solution: 10 µg/ml PI (Sigma or Molecular Probes), 0.2 mg/ml RNase (Sigma, cat. no. R-4875), 0.1% Triton X-100 in PBS

Method

1. Prepare aliquots of $0.5-2 \times 10^6$ fresh or frozen/thawed cells suspended in 100µl of PBS.

2. Add 500 µl of lysing buffer. Incubate for 15 minutes on ice.

3. Add 3 ml of 100% ice-cold methanol and mix carefully for 10 min. Store at $-20°C$ until staining, then wash once in PBS.

4. Add 100 µl of FITC-conjugated monoclonal PC-10 antibody diluted 1:20 in lysing buffer, or the equivalent amount of isotype control antibody. Incubate at room temperature for 30 min. Wash once in PBS.

5. Add 200 µl of DNA-staining solution. Incubate at room temperature for at least 15 min.

[a] Modified from the methods given in ref. 42.

4.11 Oncogene-encoded antigens

Analysis of deranged control of cell proliferation, as indicated by the expression levels of oncogenes and–anti-oncogenes (tumour-suppressor genes), is important. The gene products of many oncogenes and–anti-oncogenes can be quantified by flow cytometry, e.g. p53 and proteins of c-erbB-2 (Her-2/neu), c-myc, c-fos, types of ras, and from the retinoblastoma (pRB) and Bcl-2 gene families (15, 18). p53 is assumed to play an important role as a key regulator in control of the switch between the cell-proliferation and cell-death pathways. A series of antibodies against epitopes of mutated and wild-type forms have been used in flow cytometry for measuring p53—either in combination with measurement of differentiation markers (18) or DNA content—to characterize DNA aneuploid tumour subpopulations (24, 44).

4.12 Viral antigens

A series of viral proteins can be detected by flow cytometry, e.g. EBV, p63 and viral capsid antigen, adenovirus E1A, herpes simplex type 2, cytomegalovirus α-antigen, and the simian virus 40 (SV40) T- and V-antigens (2, 76).

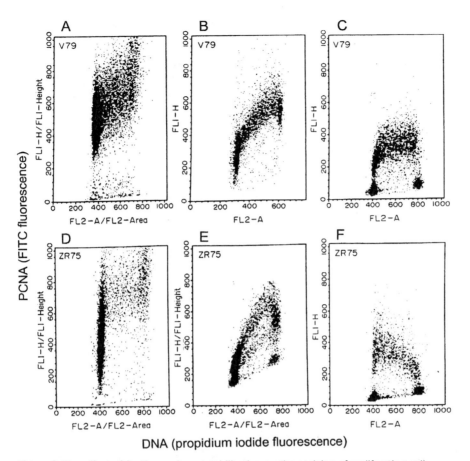

DNA (propidium iodide fluorescence)

Figure 6 The effect of fixation and permeabilization on the staining of proliferating cell nuclear antigen (PCNA) in cell cultures of V79 379A Chinese hamster fibroblasts (A–C) and ZR75 human breast tumour cells (D–F). All cycling cells were PCNA-positive after fixation in absolute methanol at −20°C (A and D). Only S-phase cells were positive after nuclear extraction with 0.25% detergent Nonidet P-40 for 10 min on ice, followed by fixation in methanol (C and F). An intermediate situation resulted from non-detergent nuclear extraction, by fixation in ice-cold 70% ethanol followed by hydrolysis in 0.1 M HCl for 10 min at 37 °C (B and E). (Reproduced from ref. 78, from Figures 1, 3, and 5, with permission from the publisher and the authors.)

References

1. Giloh, H. (1993). In *Flow cytometry—new developments* (ed. A. Jacquemin-Sablon), p. 65. Springer-Verlag, Berlin.
2. Schimenti, K. J. and Jacobberger, J. W. (1992). *Cytometry*, **13**, 48.
3. van Erp, P. E., Rijzewijk, J. J., Boezeman, J. B., Leenders, J., de Mare, S., Schalkwijk, J., van de Kerkhof, P. C., Ramaekers, F. C., and Bauer, F. W. (1989). *Am. J. Pathol.*, **135**, 865.
4. van der Linden, J. C., Herman, C. J., Boenders, J. G., van de Sandt, M. M., and Lindeman, J. (1992). *Cytometry*, **13**, 163.
5. Juan, G., Gong, J., Traganos, F., and Darzynkiewicz, Z. (1996). *Cell Prolif.*, **29**, 259.

6. Remvikos, Y., VuHai, M., Laine Bidron, C., Jollivet, A., and Magdelenat, H. (1991). *Cytometry*, **12**, 157.

7. Jacobberger, J. W. (1991). In *Methods: a companion to Methods in enzymology*, Vol. 2 No. 3 (ed. R. A. Diamond and E. V. Rothenberg), p. 207. Academic Press, San Diego.

8. Landberg, G. and Roos, G. (1992). *Cytometry*, **13**, 230.

9. Larsen, J. K., Christensen, I. J., Christiansen, J., and Mortensen, B. T. (1991). *Cytometry*, **12**, 429.

10. Jacob, M. C., Favre, M., and Bensa, J. C. (1991). *Cytometry*, **12**, 550.

11. Bauer, K. D. and Jacobberger, J. W. (1994). In *Methods in cell biology*, Vol. 41 (ed. Z. Darnzynkiewicz, J. P. Robinson and H. A. Crissman), p. 351. Academic Press, San Diego.

12. Anderson, P., Blue, M. L., O'Brien, C., and Schlossman, S. F. (1989). *J. Immunol.*, **143**, 1899.

13. Hallden, G., Andersson, U., Hed, J., and Johansson, S. G. (1989). *J. Immunol. Meth.*, **124**, 103.

14. Carayon, P. and Bord, A. (1992). *J. Immunol. Meth.*, **147**, 225.

15. Dent, G. A., Leglise, M. C., Pryzwansky, K. B., and Ross, D. W. (1989). *Cytometry*, **10**, 192.

16. Bardales, R. H., Al Katib, A. M., Carrato, A., and Koziner, B. (1989). *J. Histochem. Cytochem.*, **37**, 83.

17. Drach, J., Gattringer, C., Glassl, H., Schwarting, R., Stein, H., and Huber, H. (1989). *Cytometry*, **10**, 743.

18. Pope, B., Brown, R., Luo, X. F., Gibson, J., and Joshua, D. (1997). *Leuk. Lymphoma*, **25**, 545.

19. Turner, J. R., Tartakoff, A. M., and Berger, M. (1988). *J. Biol. Chem.*, **263**, 4914.

20. Lan, H. Y., Hutchinson, P., Tesch, G. H., Mu, W., and Atkins, R. C. (1996). *J. Immunol. Meth.*, **190**, 1.

21. Millard, I., Degrave, E., Philippe, M., and Gala, J. L. (1998). *Clin. Chem.*, **44**, 2320.

22. Redkar, A. A. and Krishan, A. (1999). *Cytometry*, **38**, 61.

23. Labalette Houache, M., Torpier, G., Capron, A., and Dessaint, J. P. (1991). *J. Immunol. Meth.*, **138**, 143.

24. Morkve, O., Halvorsen, O. J., Skjaerven, R., Stangeland, L., Gulsvik, A., and Laerum, O. D. (1993). *Anticancer Res.*, **13**, 571.

25. Lanza, F., Latorraca, A., Moretti, S., Castagnari, B., Ferrari, L., and Castoldi, G. (1997). *Cytometry*, **30**, 134.

26. Van Lochem, E. G., Groeneveld, K., Te Marvelde, J. G., Van den Beemd, M. W., Hooijkaas, H., and Van Dongen, J. J. (1997). *Leukaemia*, **11**, 2208.

27. Groeneveld, K., Te, M. J., Van den Beemd, M. W., Hooijkaas, H., and Van, D. J. (1996). *Leukaemia*, **10**, 1383.

28. Kappelmayer, J., Gratama, J. W., Karászi, É., Menendez, P., Ciudad, J., Rivas, R., and Orfao, A. *Work of the European Working Group on Clinical Cell Analysis supported by the European Community (Concerted Action BMH4-CT97–2611).*

29. Assenmacher, M., Lohning, M., Scheffold, A., Manz, R. A., Schmitz, J., and Radbruch, A. (1998). *Eur. J. Immunol.*, **28**, 1534.

30. Scheffold, A., Miltenyi, S., and Radbruch, A. (1995). *Immunotechnology*, **1**, 127.

31. Lemons, R., Forster, S., and Thoene, J. (1988). *Anal. Biochem.*, **172**, 219.

32. Ahnert-Hilger, G., Mach, W., Fohr, K. J., and Gratzl, M. (1989). *Methods in cell biology*, Vol. 31 (ed. A. Tartakoff), p. 63. Academic Press, San Diego.

33. O'Gorman, M. R. and Thomas, J. (1999). *Cytometry*, **38**, 78.

34. Knapp, W., Strobl, H., and Majdic, O. (1994). *Cytometry*, **18**, 187.

35. Prussin, C. and Metcalfe, D. D. (1995). *J. Immunol. Meth.*, **188**, 117.

36. Maino, V. C. and Picker, L. J. (1998). *Cytometry*, **34**, 207.

37. Drach, J., Gattringer, C., and Huber, H. (1991). *Br. J. Haematol.*, **77**, 37.

38. Drach, J., Gattringer, C., Glassl, H., Drach, D., and Huber, H. (1992). *Hematol. Oncol.*, **10**, 125.

39. Glogovac, J. K., Porter, P. L., Banker, D. E., and Rabinovitch, P. S. (1996). *Cytometry*, **24**, 260.

40. Darzynkiewicz, Z., Gong, J., Juan, G., Ardelt, B., and Traganos, F. (1996). *Cytometry*, **25**, 1.

41. Gong, J., Bhatia, U., Traganos, F., and Darzynkiewicz, Z. (1995). *Leukaemia*, **9**, 893.

42. Landberg, G. and Roos, G. (1991). *Cancer Res.*, **51**, 4570.

43. Wilson, G. D., Camplejohn, R. S., Martindale, C. A., Brock, A., Lane, D. P., and Barnes, D. M. (1992). *Eur. J. Cancer*, **28A**, 2010.

44. Remvikos, Y., Tominaga, O., Hammel, P., Laurent Puig, P., Salmon, R. J., Dutrillaux, B., and Thomas, G. (1992). *Br. J. Cancer*, **66**, 758.

45. Schutte, B., Scheres, H. M., de Goeij, A. F., Rousch, M. J., Blijham, G. H., Bosman, F. T., and Ramaekers, F. C. (1992). *Prog. Histochem. Cytochem.*, **26**, 68.

46. Schutte, B., Tinnemans, M. M., Pijpers, G. F., Lenders, M. H., and Ramaekers, F. C. (1995). *Cytometry*, **21**, 177.

47. Landberg, G., Tan, E. M., and Roos, G. (1990). *Exp. Cell Res.*, **187**, 111.

48. Corver, W. E., Fleuren, G. J., and Cornelisse, C. J. (1997). *Cytometry*, **28**, 329.

49. Mehta, B. A. and Maino, V. C. (1997). *J. Immunol. Meth.*, **208**, 49.

50. Faretta, M., Bergamaschi, D., Taverna, S., Ronzoni, S., Pantarotto, M., Mascellani, E., Cappella, P., Ubezio, P., and Erba, E. (1998). *Cytometry*, **31**, 53.

51. Li, X. and Darzynkiewicz, Z. (1995). *Cell Prolif.*, **28**, 571.

52. Boutonnat, J., Bonnefoix, T., Mousseau, M., Seigneurin, D., and Ronot, X. (1998). *Anticancer Res.*, **18**, 2993.

53. Jung, S. S., Nalbantoglu, J., and Cashman, N. R. (1996). *J. Neurosci. Res.*, **46**, 336.

54. Carter, L. L. and Swain, S. L. (1997). *Curr. Opin. Immunol.*, **9**, 177.

55. Yang, Y. H., Hutchinson, P., Santos, L. L., and Morand, E. F. (1998). *Clin. Exp. Immunol.*, **111**, 117.

56. Stokke, T., Erikstein, B., Holte, H., Funderud, S., and Steen, H. B. (1991). *Mol. Cell. Biol.*, **11**, 3384.

57. Mann, G. J., Dyne, M., and Musgrove, E. A. (1987). *Cytometry*, **8**, 509.

58. Neubauer, A., Serke, S., Siegert, W., Kroll, W., Musch, R., and Huhn, D. (1989). *Br. J. Haematol.*, **72**, 492.

59. Prosperi, E., Negri, C., Bottiroli, G., and Astaldi, R. G. (1996). *Anal. Cell. Pathol.*, **10**, 137.

60. Berki, T., Kumanovics, G., Kumanovics, A., Falus, A., Ujhelyi, E., and Nemeth, P. (1998). *J. Immunol. Meth.*, **214**, 19.

61. Redkar, A. A. and Krishan, A. (1999). *Cytometry*, **38**, 61.

62. Pimpaneau, V., Midoux, P., Monsigny, M., and Roche, A. C. (1991). *Carbohydr. Res.*, **213**, 95.

63. DeMaria, M. A., Zheng, Y. L., Zhen, D., Weinschenk, N. M., Vadnais, T. J., and Bianchi, D. W. (1996). *Cytometry*, **25**, 37.

64. Pollice, A. A., McCoy, J. P. J., Shackney, S. E., Smith, C. A., Agarwal, J., Burholt, D. R., Janocko, L. E., Hornicek, F. J., Singh, S. G., and Hartsock, R. J. (1992). *Cytometry*, **13**, 432.

65. Glogovac, J. K., Porter, P. L., Banker, D. E., and Rabinovitch, P. S. (1996). *Cytometry*, **24**, 260.

66. Darzynkiewicz, Z., Gong, J., Juan, G., Ardelt, B., and Traganos, F. (1996). *Cytometry*, **25**, 1.

67. Sramkoski, R. M., Wormsley, S. W., Bolton, W. E., Crumpler, D. C., and Jacobberger, J. W. (1999). *Cytometry*, **35**, 274.

68. Lukas, J., Bartkova, J., Welcker, M., Petersen, O. W., Peters, G., Strauss, M., and Bartek, J. (1995). *Oncogene*, **10**, 2125.

69. Di Vinci, A., Elio, G., Pfeffer, U., Vidali, G., and Giaretti, W. (1993). *Cytometry*, **14**, 421.

70. Gerdes, J., Li, L., Schlueter, C., Duchrow, M., Wohlenberg, C., Gerlach, C., Stahmer, I., Kloth, S., Brandt, E., and Flad, H. D. (1991). *Am. J. Pathol.*, **138**, 867.

71. Leers, M. P., Theunissen, P. H., Ramaekers, F. C., and Schutte, B. (1997). *Cytometry*, **27**, 283.

72. Larsen, J. K., Christensen, I. J., Christiansen, J., and Mortensen, B. T. (1991). *Cytometry*, **12**, 429.

73. Bravo, R. and Macdonald-Bravo, H. (1987). *J. Cell Biol.*, **105**, 1549.

74. Kurki, P., Ogata, K., and Tan, E. M. (1988). *J. Immunol. Meth.*, **109**, 49.

75. Bachelet, M., Mariethoz, E., Banzet, N., Souil, E., Pinot, F., Polla, C. Z., Durand, P., Bouchaert, I., and Polla, B. S. (1998). *Cell Stress Chaperones*, **3**, 168.

76. Sladek, T. L. and Jacobberger, J. W. (1993). *Cytometry*, **14**, 23.

77. Gong, J., Traganos, F., and Darzynkiewicz, Z. (1993). *Cancer Res.*, **53**, 5096.

78. Wilson, G. D., Camplejohn, R. S., Martindale, C. A., Brock, A., Lane, D. P., and Barnes, D. M. (1992). *Eur. J. Cancer*, **28A**, 2010.

79. Francis, C. and Connelly, M. C. (1996). *Cytometry*, **25**, 58.

Chapter 10

Analysis of DNA—measurement of cell kinetics by the bromodeoxyuridine/anti-bromodeoxyuridine method

George D. Wilson

Gray Laboratory Cancer Research Trust, PO Box 100, Mount Vernon Hospital, Northwood HA6 2JR, UK

1. Introduction

Cell kinetics is primarily concerned with the measurement of time parameters in the growth of biological systems. This definition sets it apart from other techniques involved in the study of cellular proliferation. Most methods involve a 'snapshot' assessment of cells in a particular phase of the cell cycle (e.g. mitotic index, tritiated thymidine labelling index or S phase fraction from DNA histograms) or cells actively involved in a proliferation cycle (e.g. Ki-67, PCNA staining—see Chapter 9). Whilst static parameters of cell proliferation can be informative, they give no indication of the rates of cell-cycle progression and are subject to ambiguity as will be discussed later in this chapter.

Traditional methods used to measure cell kinetics have involved the use of radioactive precursors of DNA, in particular tritiated thymidine ([^{3}H]TdR), and autoradiography to visualize the incorporation of the radioisotope into cells by formation of silver grains in a photographic emulsion layered over the tissue section on a microscope slide (1). The technique measured cell kinetics by counting the progress of cells, labelled in S phase, through the mitotic window. This technique has contributed much to our understanding of the organization of proliferation in both normal tissues and tumours. However, it suffers from the problems of being time-consuming and arduous; it is also difficult to apply to slowly proliferating tissues and its use in humans has been mainly restricted to in-vitro studies due to the ethical restraints of administering radioactive DNA precursors to patients.

Flow cytometry (FCM) has many attributes which sets it apart from other techniques, particularly its speed, quantitative power, and ability to make several measurements simultaneously. Several methods have been developed to study cell kinetics with FCM. Chapter 10 describes the use of bromodeoxyuridine

(BrdUrd) quenching of Hoechst 33342 binding to study successive cell cycles in experimental systems. In this chapter, I will be dealing with the measurement of cell kinetics, using FCM, by detecting BrdUrd incorporation using a monoclonal antibody. This technique is applicable not only to experimental systems but it also can be used in human tumours *in vivo*.

2 Basic cell-kinetic concepts

Readers are referred to the comprehensive treatise by Gordon Steel (1) for the basic theory of growing cell populations and a detailed review of cell kinetics in experimental systems and human tumours. Some of the concepts and definitions are outlined below.

2.1 Cell-cycle time (T_C)

T_C is defined as the time interval within which one cell, or a group of cells, complete a mitotic cycle, that is, from birth at mitosis to eventual splitting to form two progeny at the next mitosis. The cell cycle is made up of discrete subcompartments termed G_1, S, G_2, and M (mitosis). Each phase of the cell cycle accomplishes a sequence of biochemical events to enable the successful production of viable daughter cells. Events in G_1 prepare and control the entry of cells into DNA synthesis. During S phase, DNA is correctly duplicated, whilst in G_2 the events that allow entry into mitosis are completed. During mitosis the mitotic spindle is assembled, chromosomes become condensed, and the nuclear envelope breaks down in preparation for cell division. Clearly it is an important decision for a cell to divide, and the events required for this process are under feedback controls and scrutiny to ensure that correct replication takes place (see refs 2 and 3 for reviews). Each phase of the cell cycle will have a finite time for its completion, and thus the number of cells found in any particular phase will depend on the time they spend in that phase relative to the cell-cycle time.

2.2 Volume doubling time (T_d)

In a 3-dimensional system, such as a solid tumour, the situation is more complex than that found in experimental cells in that not all cells will be involved in an active cell cycle. This introduces two new concepts: growth fraction and cell loss. The term G_0 has been adopted to refer to those cells that are out of cycle. However, this rather nondescript term is used to describe cells that may be out of cycle for vastly different reasons (see *Figure 1*). Cells may be out of cycle because they have started the process of differentiation and maturation or they may have reached the end of their life span and are undergoing cell death. They may be in a poorly vascularized area of the tissue and cannot continue their cycle due to nutrient deprivation. This may leave them in a state of temporary hypoxia, in which they may be able to re-enter the cycle if the microenvironment improves or they may die if the situation becomes chronic. Cell loss also describes a diversity of underlying processes from exfoliation in mucosal tissues

CELL PRODUCTION   CELL LOSS

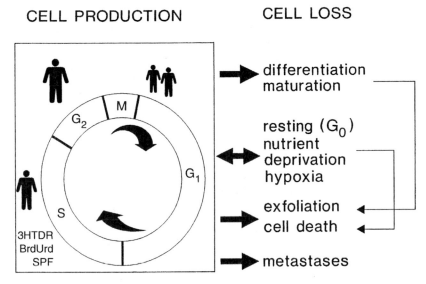

Figure 1 Schematic representation of cellular proliferation showing the balance between cell production and cell loss. SPF in the S phase fraction.

to programmed cell death or apoptosis, necrosis, or even metastatic loss in tumours. The volume doubling time represents the time interval in which a tumour doubles its volume and will be subject to cell loss, the growth fraction, as well as the influence of any host cells contained within the tumour mass.

2.3 Potential doubling time (T_{pot})

The volume doubling time is often a poor indicator of cellular proliferation. It tells nothing of the rate of cell production that may be an important determinant of treatment success. The T_{pot}, or turnover time in normal tissues, is defined as the time within which a cell population would double its number if cell loss did not occur. In normal tissues the absolute rate of production of cells is assumed to remain constant, whilst the T_{pot} of tumours assumes that cells that would have been lost remain in the population without changing the growth fraction. These parameters can be calculated knowing the duration and the number of cells within any proliferation-active, cell-cycle phase. They are usually calculated from either the mitotic index and the duration of mitosis or, more commonly, from the labelling index (*LI*), i.e. the proportion of cells which take up a DNA precursor and the duration of S phase (T_S) by the formula:

$$T_{pot} = \lambda \times T_S/LI.$$

The term λ takes into account the non-uniform distribution of age through the cell cycle and typically lies between 0.693 and 1. Thus the T_{pot} will be shorter than the T_d but longer than the T_C depending on the growth fraction. The T_{pot}, as will be discussed later, appears to be a parameter of clinical significance as

161

it measures the maximum potential for growth in a tumour system; this may help to identify patients whose tumours are capable of repopulating during or after radiation or chemotherapeutic treatment (4).

3 Background to the BrdUrd technique

The technique became possible due to the development of monoclonal anti-bodies that recognize halogenated pyrimidines incorporated into DNA (5). Dolbeare and colleagues (6) then developed a simultaneous staining method, using FCM, to study the incorporation of BrdUrd relative to DNA content measured by propidium iodide (PI). The general approach to measuring cell kinetics is to identify a 'window' in the cell cycle and measure the movement of a cohort of labelled cells through the window. With [^{3}H]TdR, the only identi-fiable window is mitosis, consequently, the per cent labelled mitosis (p.l.m.) analysis was developed. However, with the BrdUrd/PI method windows can be set in any phase of the cell cycle (see *Figure 2*) looking at either the BrdUrd-labelled or -unlabelled population using appropriate computer-generated regions.

The essence of the procedure is to pulse-label with BrdUrd by a short incubation *in vitro* or by a single injection *in vivo*; samples are then taken at time intervals thereafter and stained after fixation in ethanol. The cells are stained using a monoclonal antibody against BrdUrd that can either be directly con-jugated to a fluorochrome (usually FITC) or alternatively bound to a second antibody conjugated with FITC. The cells are then counterstained with PI to measure DNA content and analysed on the flow cytometer for red (DNA) and green (BrdUrd) fluorescence. The results are displayed as red (x-axis) vs. green (y-axis) bivariate distributions.

Figure 2 shows a series of such distributions obtained at hourly intervals after pulse-labelling V79 Chinese hamster fibroblasts with 10 μM BrdUrd for 20 min *in vitro*. In the profile recorded after 1 h, the BrdUrd-labelled, S-phase cells are clearly identified by their green fluorescence and lie between the G_1 and G_2 populations. The latter two populations can also be separately identified by virtue of their lack of BrdUrd uptake and the difference in their DNA content. The BrdUrd labelling is classically crescent-shaped with lower levels of uptake in cells that have just entered S phase from G_1 and those found in late S phase about to enter G_2. With time, the profiles show changing patterns. At 2 h, the labelled cohort has become slightly skewed to the right as the cells have made more DNA. One hour later sees the appearance of BrdUrd-labelled cells in G_1 (see arrow), these represent cells in late S at the time of labelling which have finished S phase, gone through G_2, divided and become two daughter G_1 cells. It should also be noted that the original G_2 population has also disappeared, as these cells will divide prior to the labelled cells. As time progresses the BrdUrd-labelled cohort moves through S phase and more cells divide. It is also evident that the original G_1 population moves into S phase; these cells can be identified by virtue of their lack of BrdUrd but they have S phase DNA (see arrow in the

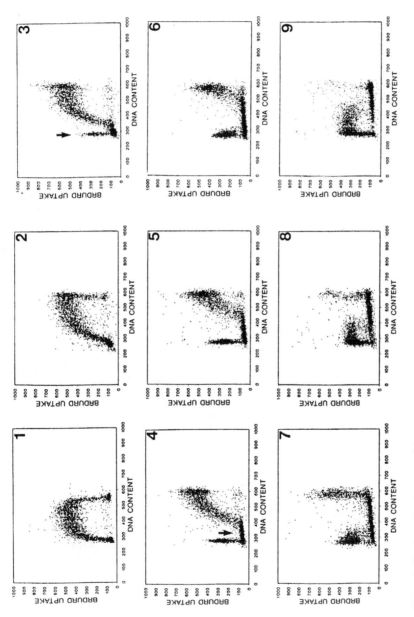

Figure 2 Bivariate distributions of BrdUrd incorporation (y-axis) vs. DNA content (x- axis) obtained from V79 cells pulse-labelled with BrdUrd and sampled and stained at hourly intervals after removal of the DNA precursor. The numbers on the cytograms represent the number of hours after the pulse-label.

163

4 h profile). This is another attribute of the technique, as each population can be followed as it moves through its cell cycle whether it is BrdUrd-labelled or not. At 6 h some of the labelled cells in G_1 begin to move into S phase for the second time as they complete their transit of the G_1 phase. This is more evident in the subsequent profiles. Eventually the staining profile will revert to that seen in the 1-hour profile when all cells have completed their cycle; this would probably take 12 h in this particular example.

4 Methods of BrdUrd incorporation

4.1 *In-vitro* incorporation

Many studies have been made involving the incorporation of [³H]TdR into cells, tissues, and tumours and this methodology is applicable to BrdUrd. Incorporation into cell cultures is usually achieved by incubation with 10–20 μM BrdUrd for 20–30 min; this will achieve a pulse-label. The drug would normally be made up as a stock solution of 1 mM in the medium used for the particular cell line. The BrdUrd solution should always be sonicated or shaken in warmed medium to ensure that it is fully dissolved. The stock can be diluted to the required final concentration by addition to the cell culture. It is important to make sure everything is kept at 37 °C as small fluctuations in temperature can cause cell-cycle perturbations. This is particularly required at the next stage where the BrdUrd is washed out after the labelling period. At least two washes in warm medium are required. It is essential that there is no contaminating BrdUrd left as it will continue to be incorporated as cells move from G_1 into S. It is imperative that control cultures should be sham-treated in experiments involving cell-cycle blocking agents. After removal of the DNA precursor, fresh medium is added and the cells allowed to progress through their cycle until the required sampling time. In continuous labelling studies, the BrdUrd is usually incubated for a complete cell-cycle generation until all cells become labelled. This is particularly useful for looking at the recruitment of cells into the cycle. Care must be taken to ensure that the concentration of BrdUrd used does not have a cytostatic effect. This can be done by comparing the cellular doubling time by counting cells in non-labelled cultures and labelled cultures.

Incorporation into solid tissue such as a tumour presents some problems. There are two approaches, the first would be to produce a single-cell suspension from the tumour using enzyme dissociation and to label this as above. The second approach would be to label tissue fragments or slices and to produce a cell suspension afterwards for FCM. When labelling solid tissue it should be noted that there would be no cell-cycle progression, as cells will incorporate DNA precursors but not continue to traverse their cell cycle due to the trauma of removal from their environment *in situ*. Thus these experiments are only used to obtain an LI. In this respect, the production of a cell suspension prior to labelling can be subject to cell selection and cellular damage. It is intuitively better to label intact tissue pieces first and to produce a cell suspension later.

Protocol 1 describes a method we have used to good effect for labelling tissue fragments (7); this was developed from that described by Steel and Bensted (8) for [³H]TdR and autoradiography.

Protocol 1

In-vitro incorporation of BrdUrd into solid tissues

Equipment and reagents

- Polypropylene LP/3 tubes[a]
- Tube caps and rubber gaskets, each with a centrally pierced hole
- Metal retaining tube constructed as described in ref. 8
- Gelatin capsules
- Tissue culture medium[b] with 10% FCS, 20 mM Hepes buffer (pH 7.4), 50 μM BrdUrd (Sigma)
- Catalase (Sigma)
- 30% hydrogen peroxide (Sigma)
- PBS, pH 7.4

Method

1. Cut fresh tumour or tissue pieces into 1–2 mm³ fragments using a scalpel.

2. Add up to 10 of these fragments to a polypropylene LP/3 tube[a] containing 2.5 ml of culture medium with 10% FCS, 20 mM Hepes buffer, 50 μM BrdUrd. Seal the tube with a cap through which a central hole has been pierced.

3. Place a rubber gasket (with central hole) on top of the cap and put the tube into a metal retaining tube.

4. Prepare a similar LP/3 tube with 2.5 ml of medium with a gelatin capsule in which there is a small amount of solid catalase.

5. Add three drops of 30% hydrogen peroxide and then cap the tube (also with a central hole).

6. Place in the metal retaining tube on top of the other LP/3 tube (cap to cap) and seal the metal tube with a screw top.[c]

7. Incubate at 37°C for 1 hour with constant shaking or rotation.

8. At the end of the incubation, release the pressure under water. Pool the tumour pieces and wash in PBS prior to cell or nuclei dissociation.

[a] It is essential that polypropylene tubes are used as pressures of up to 10 atmospheres are generated by the oxygen produced from the catalase and H_2O_2.

[b] The choice of medium is unimportant; we use α-modified minimum essential medium.

[c] Step 6 needs to be carried out quickly as the generation of O_2 is rapid. The presence of high-pressure oxygen reduces the likelihood that anoxia will develop in the fragments which will inhibit the uptake of BrdUrd. It also increases the distance the DNA precursor can diffuse, which is typically 150 μm under ambient conditions; this ensures that all cells will be labelled.

4.2 *In-vivo* incorporation

The half-life of BrdUrd in rodent and human blood is short, typically 10–15 min, such that a single intraperitoneal (i.p.) or intravenous (i.v.) injection acts as a pulse-label. In experimental animals the BrdUrd should be made up in 0.9% saline immediately prior to use, again taking care to ensure that the drug is in solution. An i.p. injection of 50–100 mg/kg is sufficient for pulse-labelling; this would normally be made as a 10 mg/ml solution such that 0.05–0.1 ml/10 g is administered to the animal. In mice, as many as eight doses of 100 mg/kg have been given over a 24-hour period in continuous labelling studies.

In man, several grams of BrdUrd have been given as a continuous infusion when it has been used as a radiosensitizer. A much lower dose is required for cell kinetic studies; we administer 200 mg in 20 ml of saline as a single intravenous bolus injection over a few min. Other groups give 100 mg/m^2, whilst 500 mg have also been used without any adverse toxicity. To date, over 2000 patients have been given BrdUrd or iododeoxyuridine (IdUrd) without any reports of acute or long-term toxicity.

5 Tissue preparation

A prerequisite for flow cytometric analysis is the preparation of single cells or nuclei from cell cultures or solid tissues. Chapter 3 deals with preparative techniques in a variety of circumstances. I will restrict this section to discussion of the pepsin digestion method for obtaining nuclei.

The objective of any disaggregation procedure is to obtain the maximum number of intact cells or nuclei from a given amount of tissue, yielding a representative sample of the original population. Production of cell suspensions by either mechanical or enzymatic methods does have several limitations, particularly in the clinical setting.

(a) One limitation is the yield from solid tumours, particularly if only biopsy material is available; at least 10^6 cells are required for BrdUrd staining.

(b) In addition, production of good cell suspensions should be done promptly after the surgical procedures. This means that the specimen needs to be collected from theatre and there must be laboratory facilities on site or close by.

(c) When dealing with tumours from different body areas, it is not uncommon to use quite different procedures to produce a cell suspension.

(d) Retention of the cytoplasm is often the cause of much of the non-specific staining associated with primary and secondary antibodies.

Although cell suspensions are a perquisite for any surface or cytoplasmic labelling, they are not necessary for staining procedures restricted to nuclear elements. Production of nuclei suspensions with the pepsin digestion procedure outlined in *Protocol 2* overcomes the shortcomings listed above. The yield is considerably improved with pepsin, which means that only small amounts of tissue (<20 mg) are required. The handling procedures are minimal as the biopsy is simply fixed in 70% ethanol and can be stored at 4°C until staining or sent by post to a centre with an FCM facility. The method is amenable to all tumour types and so procedures can be standardized. Non-specific staining is greatly reduced and the coefficient of variation (CV) of the DNA profile is often reduced compared to intact cells.

Protocol 2

Procedure for obtaining nuclei from ethanol-fixed tumours

Equipment and reagents

- Pepsin solution: 0.4 mg/ml pepsin (Sigma) in 0.1 M HCl. Dissolve the required amount of pepsin in 1–2 ml of PBS or distilled water then add the HCl. If the solution looks cloudy warm to 37°C.
- 35 μm nylon mesh (Lockertex or Small Parts)

Method

1. Mince fresh tumour specimens with scissors or leave as large pieces.
2. Add 10 ml of 70% ethanol and store at 4°C. Fix overnight or longer prior to staining.
3. Decant the ethanol and mince the tissue into small (1 mm) fragments.
4. Add 5–10 ml of the pepsin solution.
5. Incubate on a rotor or shaking water bath for 30–60 min at 37°C depending on the tissue.
6. Further dissociate by pipetting.
7. Filter through the nylon mesh into a conical-bottomed 10 ml tube.
8. If required, count the suspension and remove the desired number of nuclei.
9. Centrifuge at 700 g for 5 min. Resuspend the pellet in the acid required for DNA denaturation (see Protocol 3, step 1).

6 Staining procedures

Several procedures have been described for identifying cells that have incorporated BrdUrd. The basic immunochemical staining with monoclonal antibody, either directly conjugated to FITC or indirectly through a second antibody, varies little between procedures. Different monoclonals have different affinities for the halogenated pyrimidines and this can be exploited in double-staining procedures (see Section 6.2). The diversity of techniques arises from the requirement for partial DNA denaturation to permit access of the monoclonal antibody to its epitope on the incorporated BrdUrd.

6.1 DNA denaturation

The DNA denaturation must be rigorously controlled to achieve sensitive detection of BrdUrd without disruption of the binding stoichiometry of the

167

intercalating dye, PI, which requires double-stranded DNA. The techniques that will be described include the use of acid, the use of heat, and the use of endonuclease/exonuclease enzymes. Each procedure has its advantages and disadvantages. The HCl method is relatively mild and applicable to all cell types; its disadvantage is that the level of denaturation achieved without disruption of the PI staining may not be enough to permit sensitive detection of BrdUrd incorporated at low levels. This has been improved by Schutte *et al.* (9) who combined acid denaturation with pepsin enucleation of cell suspensions.

The thermal denaturation procedures in formamide or water greatly increase the sensitivity of BrdUrd detection. Beisker *et al.* (10) reported that the fluorescence ratio of BrdUrd labelled S-phase cells to unlabelled G_1 cells was fivefold using standard 2 M HCl at 25°C, but could be increased to 150-fold by extracting histones with 0.1 M HCl, 0.7% Triton X-100 followed by denaturation in water for 10 min at 100°C. The major drawback of thermal denaturation techniques is that substantial cell loss can occur; up to 90% of haematopoietic cells can be lost during the process. This renders the technique unsuitable for studies where the sample is at a premium, such as those with human tumours. In experimental systems where material is not limiting, the thermal techniques would be particularly applicable for studies of drug resistance, unscheduled DNA synthesis, etc.

Combining BrdUrd staining with surface markers or other labile antigens has proven difficult due to the nature of the denaturation procedures. Perhaps the most promising method to obtain sensitive detection of BrdUrd without severe cell loss, morphological change, or protein loss is to use a restriction endonuclease and exonuclease (11). The disadvantage of this method is the extra cost and also that different cell types might require different endonucleases for optimal detection conditions.

Protocol 3

Denaturation procedures for BrdUrd detection

Reagents

For Part A
- 2 M HCl
- PBS

For Part B
- 2 M HCl
- PBS
- Pepsin (Sigma)

For Part C
- PBS
- RNase (Sigma)
- 0.1 M HCl
- Triton X-100
- 50% formamide

- Ice-bath

For Part D
- PBS
- 0.1 M citric acid
- Triton X-100
- Tris buffer: 0.1 M Tris–HCl, 50 mM NaCl, 10 mM $MgCl_2$ pH 7.5
- *Bam*H1 buffer: 10 mM Tris–HCl, 5 mM $MgCl_2$, 1 mM mercaptoethanol pH 8.0
- *Bam*H1 (Sigma)
- Exonuclease III buffer: 66 mM Tris–HCl, 0.66 mM $MgCl_2$, 1 mM mercaptoethanol pH 7.6
- Exonuclease III (Sigma)

A. HCl acid alone

1. Resuspend $1–3 \times 10^6$ ethanol-fixed cells or nuclei from *Protocol 2*, step 9 in 2.5 ml of 2 M HCl at room temperature. Incubate nuclei for 10–12 min and intact cells for 20–30 min, both with occasional mixing. Add 5 ml of PBS.

2. Centrifuge at 700 g for nuclei and 400 g for cells for 5 min at room temperature.

3. Decant the supernatant and repeat the washing to get rid of all the acid.

4. Alternatively, resuspend the pellet in 5 ml of 0.1 M sodium tetraborate to neutralize the acid and wash twice in PBS.

B. Pepsin/HCl combined enucleation and denaturation

1. Resuspend ethanol-fixed cells in 2.5 ml of 2 M HCl containing 0.2 mg/ml of pepsin. Dissolve the pepsin in a small amount of water or PBS first. Incubate at room temperature for 20 min (this may vary according to the cell type).

2. Wash twice with 5 ml PBS, centrifuging for 5 min at room temperature at 700 g between washes.

C. Thermal denaturation methods

1. Incubate ethanol-fixed cells in 1.5 ml of PBS containing 1 mg/ml RNase for 20 min at 37 °C.

2. Centrifuge at 400 g for 5 min at room temperature.

3. Resuspend the pellet in 2 ml of 0.1 M HCl containing 0.7% Triton X-100 for 10 min at 0 °C.

4. Centrifuge for 5 min at room temperature at 400 g and wash with 5 ml of cold distilled water.

5. Resuspend in 1.5 ml of 50% formamide for 30 min at 80 °C or 2 ml of distilled water for 10 min at 100 °C.

6. Plunge the tubes into an ice-bath to stop the reaction.

7. Wash with 5 ml PBS.

D. Endonuclease/exonuclease denaturation

1. Incubate ethanol-fixed cells in 1 ml of PBS containing 1 mg/ml RNase for 30 min at 37 °C.

2. Centrifuge (5 min, 700 g, room temperature) and wash in 5 ml PBS.

3. Centrifuge (5 min, room temperature) and resuspend the pellet in 1 ml of 0.1 M citric acid containing 0.5% Triton X-100 and incubate for 10 min on ice.

4. Centrifuge (5 min, 700 g, room temperature) and wash with 3 ml of the Tris buffer.

5. Centrifuge (5 min, 700 g, room temperature) and resuspend the pellet in 100 μl of *Bam*H1 buffer with 60 units of *Bam*H1. Incubate for 30 min at 37 °C.

6. Centrifuge (5 min, 700 g, room temperature) and wash with 3 ml of the Tris buffer.

7. Resuspend the pellet in 100 μl of exonuclease III buffer with 100 units of exonuclease III. Incubate for 30 min at 37 °C.

8. Wash with the Tris buffer and then with PBS.

6.2 Antibody staining

There are many commercial sources of monoclonal antibodies recognizing the halogenated pyrimidines. The choice of antibody often depends on individual preference for one or other company. However, several points should be taken into account as the quality of the measurement will depend on the purity, affinity, and specificity of the antibody. For instance, high-affinity antibodies such as IU-4 (Caltag) will be required if low levels of BrdUrd are to be detected. Most antibodies are divalent and their binding is limited when BrdUrd substitution is less than 1%. In some instances it may be preferable to use a rat-derived monoclonal such as BU/175 (Harlan Sera-Lab) to reduce non-specific staining if working with mouse tissue, or to combine with another mouse-derived monoclonal. Different antibodies have different specificities for the various halogenated pyrimidines and this can be exploited in double-labelling techniques (see *Protocol 5*). *Table 10.1* lists the cross-specificity of some antibodies to the different DNA precursors.

Table 1 Specificity of commercial antibodies for different halogenated pyrimidines

Antibody	BrdUrd	IdUrd	CldUrd
Br-3 (Caltag)	+++	–	++
IU-4 (Caltag)	+++	+++	+++
B44 (Becton Dickinson)	+++	+++	–[a]
BU/175 (Sera-Lab)	+++	–	++

[a] This result is only obtained after two 30 min washes in a Tris buffer with a high salt concentration (12).

As with all antibody staining the optimal working concentration should be assessed by dilution analysis and this may vary from batch to batch. In double-staining techniques, staining can be optimized by using a high concentration of the first monoclonal antibody (e.g. Bu/175) for CldUrd such that all possible binding sites are occupied, and then using a lower concentration of the second monoclonal (e.g. B44) for IdUrd thereby increasing its preferential binding to the more abundant IdUrd sites.

The following protocol is our standard immunochemical detection system for BrdUrd utilizing the rat anti-BrdUrd antibody (Harlan Sera-Lab), if mouse antibodies are used the second antibody should be substituted with a goat anti-mouse–FITC. We routinely use indirect fluorescence to increase the signal, as the BrdUrd levels are often low in human tumours. However, most of the major monoclonals are now available as directly labelled conjugates and these work well in experimental systems and, of course, are particularly useful in double-labelling experiments. Again, the choice of the second antibody source is not crucial, we routinely use Sigma antibodies.

Protocol 4

Immunochemical detection of BrdUrd

Reagents

- PBS/Tween/NGS: PBS containing 0.5% Tween-20, 0.5% normal goat serum (NGS— this blocking agent can be substituted with 1% BSA)
- Anti-BrdUrd antibody
- Goat anti- rat IgG (whole molecule)–FITC conjugate (Sigma)
- Propidium iodide (Sigma or Molecular Probes)

Method

1. After denaturation and washing, resuspend the pellet in 0.5 ml of PBS/Tween/NGS and 20 µl of rat anti-BrdUrd antibody. Incubate for 1 hour at room temperature with occasional mixing.
2. Add 5 ml PBS and centrifuge at 700 g for 5 min.
3. Resuspend the pellet in 0.5 ml PBS/Tween/NGS containing 20 µl of goat anti- rat IgG (whole molecule)–FITC conjugate. Incubate for 30 min at room temperature.
4. Add 5 ml of PBS and centrifuge at 700 g for 5 min.
5. Resuspend the pellet in 1–2 ml of PBS containing 10 µg/ml propidium iodide.
6. If desired, analyse the preparation immediately on the FCM.

Protocol 5

Double staining of BrdUrd and IdUrd

Reagents

- NGS
- PBS/Tween/NGS (see *Protocol 4*)
- Anti-BrdUrd antibody–FITC conjugate (Br-3, Caltag)
- Anti-IdUrd antibody (IU-4 antibody, Caltag)
- Goat anti-mouse IgG in PBS
- Goat anti-mouse IgG R–phycoerythrin conjugate (Dako or Becton Dickinson)
- 7-aminoactinomycin D (Sigma)

Method

1. After denaturation and washing resuspend the pellet in 100 µl of PBS/Tween/NGS containing 10 µl of Br-3–FITC conjugate. Incubate for 1 h at room temperature.
2. Add 5 ml of PBS and centrifuge at 700 g for 5 min.
3. Block unoccupied binding sites by incubating first in 10% NGS for 15 min followed by a 1:20 dilution of goat anti-mouse IgG in PBS for 15 min.
4. Wash in PBS and resuspend the pellet in 100 µl of PBS/Tween/NGS containing 5 µl of IU-4 antibody. Incubate for 1 h at room temperature.

Protocol 5 continued

5. Add 5 ml of PBS and centrifuge at 700 g for 5 min.

6. Resuspend the pellet in 100 µl of PBS/Tween/NGS containing 10 µl of goat anti-mouse IgG R–phycoerythrin conjugate. Incubate for 30 min at room temperature.

7. Add 5 ml of PBS and centrifuge at 700 g for 5 min.

8. Resuspend the pellet in 1 ml of PBS containing 10 µg/ml of 7-aminoactinomycin D[a]

9. Analyse on the FCM.

[a] 7-AAD has less spectral overlap with R-PE than PI, see Chapter 2.

7 Flow cytometry

This type of staining can be analysed on any of the modern flow cytometers, with the proviso that the machine is equipped with a pulse-processing facility to enable the discrimination of cell doublets (see Chapter 6). When using the Becton Dickinson FACScan, the propidium iodide should be collected in FL3 rather than FL2 to overcome any crossover with the FITC in FL1. Routinely, the FL3 detector should be set at around 400V whilst FL1 is usually around 500V. The FITC signal can be collected with log or linear amplification. Controls, either without BrdUrd or the monoclonal antibody, should be included wherever possible to determine the lower limit of detection of the DNA precursor in the bivariate profile. Controls are particularly important in double staining to determine the level of compensation to be set.

8 Examples of data and analysis

8.1 Experimental systems

Figure 2 shows the type of data that can be obtained from cells grown in culture and incubated with BrdUrd in a pulse-labelling experiment. Many different experiments can be designed using this type of approach to study progression through the cell cycle.

Figure 3 shows an example of an experiment where two cell-cycle perturbing drugs, hydroxyurea and 4-hydroxyanisole, were compared. The cell line was the V79 cells described earlier. The experimental design was to pulse-label with 10 µM BrdUrd for 30 min, followed by washout. The cell cultures were then allowed to progress through their cell cycle for 3 h before the drugs were added; both at 0.3 mM. The drugs were then incubated with the cells for 5 h before being washed out. Samples were taken at hourly intervals through the whole experiment up to 16 h.

A total of six different analysis windows were set in G_1, G_2, and a narrow window in mid-S looking at both the BrdUrd-labelled and the non-labelled populations. In this way the effects of the drugs could be studied on each cell-cycle phase. For instance, the G_1 window (see *Figure 3A*) shows clearly that the

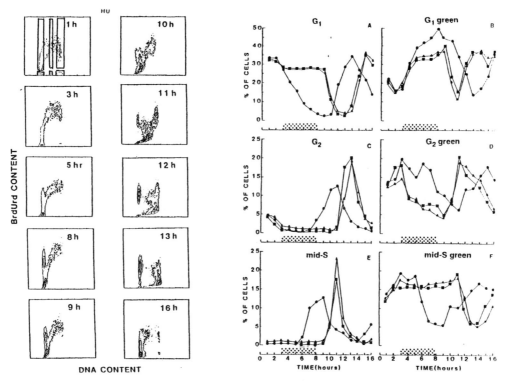

Figure 3 Experimental data obtained from untreated (●) V79 cells or treated with either 0.3 mM hydroxyurea (▼) or 4- hydroxyanisole (■). On the left are the staining profiles obtained in response to hydroxyurea; the analysis regions indicated in the 1 h profile. On the right the number of cells in these regions is plotted as a function of time. The hatched area represents the time period in which the cells were exposed to the drugs.

transit of cells from G_1 into S is completely blocked during the period with the drugs. Upon removal, there is a very rapid efflux of cells into S. This effect demonstrates the classical effect of hydroxyurea in blocking at the G_1/S interface. Whereas the G_1-labelled window (see *Figure 3B*) shows that BrdUrd-labelled cells continue to enter G_1, at least for the first 2 h in the presence of drug. The result suggests that cells in G_2 and late S are able to continue their cell-cycle progression in the presence of the drug, but that cells in early to mid-S are stopped at their cell-cycle position when the drugs are added. It is evident, particularly from the G_2 and mid-S windows (see *Figure 3C, E*), that after drug removal there is a rapid burst of cells through those cell-cycle phases. This effect is demonstrated by the time taken for these cohorts of cells to go through the window compared to the untreated control cultures.

These basic regions can be modified depending on the type of experiment being undertaken. In experimental tumours we have measured G_2 delay after irradiation by using the window in G_1 which measures the division rate of BrdUrd-labelled cells (13). This analysis measures the cell-cycle delay experienced by cells irradiated in S phase as they attempt to repair the damage inflicted by ionizing radiation. Equally, we can measure the delay experienced by cells in G_1 after irradiation, either by using a G_1 unlabelled window or alternatively a window in early S looking at the efflux of cells from the G_1 population.

The basic cell-cycle parameters can be calculated by a variety of analyses. The most common method to measure the T_C is to set two narrow windows in mid-S (determined by measuring the mean DNA fluorescence of the G_1 and G_2 populations). One widow is set in the BrdUrd-labelled population only, whilst the other spans both labelled and unlabelled populations. The data is expressed as the ratio of labelled to total cells in mid-S. Initially, the ratio should be 1 as all cells should be in the labelled window. This, however, may not be the case in experimental tumours, as some cells in S phase do not always incorporate the DNA precursor. This ratio should remain at a maximum for a period equal to 0.5 T_S, i.e. when the cells in early S phase at the time of labelling reach the window (see *Figure 2*). The ratio will then fall as the labelled cells clear the region followed by the original (unlabelled) G_1 population reaching mid-S. Thus the ratio remains low until the BrdUrd population complete G_2 and G_1 and re-enter S phase for the second time—at which time the ratio rises and forms a second peak. The mid-point of that second peak is used to determine T_C. However, in tumours this second peak is often 'damped' due to the variation in cell-cycle transit times and to the failure of some cells to go through a second round of division.

The duration of G_2 can be measured from the entry of labelled cells into G_1 (see *Figure 3B*). This estimate can be achieved by extrapolating a line back to zero which has been fitted to time-points after the first 2 h up until the entry plateau when all labelled cells are in G_1. Prior to 2 h, this region may contain cells that are still in early S but are included due to the spread of the G_1 population.

The duration of S phase can be calculated using a technique called 'relative movement' (RM) (14). This forms the basis of a method to measure human tumour cell kinetics *in vivo* and will be discussed in the next section.

8.2 Human tumours

In-vitro measurements of BrdUrd will only yield the *LI*. A real advance is the ability to measure time parameters of human tumours as these should provide better information about the proliferation characteristics of tumours, which are possibly an important factor in determining the success of radio- or chemo-therapy. These parameters can now be measured by injection of BrdUrd to patients several hours prior to biopsy or surgical resection (14, 15). Not only the *LI* but also the T_S can be measured from a single sample, and the T_{pot} computed from these as described in Section 2.3.

The procedure is based on a measurement of the mean DNA content of the BrdUrd-labelled cells and is outlined in *Figure 4*. Immediately after labelling, the mean of the BrdUrd-labelled population is approximately mid-way between G_1 and G_2, as there is a uniform distribution of cells throughout S phase. To

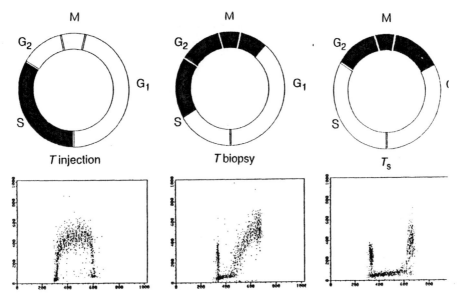

Figure 4 Calculation of *LI*, T_S, and T_{pot} from a single observation. The upper panel shows the movement of a BrdUrd label around the cell cycle. The lower panel shows the profiles that would be obtained on the FCM at those times indicated.

quantitate the function it is also necessary to measure the mean DNA contents of the G_1 and G_2 populations (this can be done from the single-parameter DNA profile). The *RM* is calculated by subtracting the mean DNA of the G_1 population from that of the labelled cells and dividing it by G_1 subtracted from G_2. If there is a uniform distribution at time zero, it should give a value of 0.5. With time, the mean value of the labelled cohort (which remains undivided) will increase as it progresses through S to G_2. If we make the assumption that the progression through S phase is linear, a point will be reached when all the labelled cells that remain undivided are in G_2. The *RM* value will equal 1.0, and this time will be equivalent to the T_S as the cells in G_2 will have been those in early S at the time of injection. Thus from a single observation, made at a known time greater than T_{G2} but less than $T_S + T_{G2}$, the T_S can be computed assuming the value of *RM* is 0.5 at time zero and 1.0 at T_S.

The *LI* can be calculated as the proportion of BrdUrd-labelled cells taking into account that some will have divided and produced two daughter cells at the time of the biopsy. A simple correction can be made to calculate what the true *LI* is by setting a box around those labelled cells which appear in the G_1 population, halving this number, and subtracting it from the total number of labelled cells and from the total cell number.

Recently, attempts have been made to put the calculation of T_{pot} on a more rigorous mathematical basis. White *et al.* (16) have introduced a quantity, *v*, which is a function of the fraction of labelled divided cells and the fraction of labelled undivided cells. This value relates T_S to T_{pot} and is only weakly dependent

175

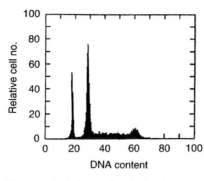 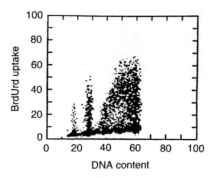

Figure 5 Staining profile obtained from a human squamous cell carcinoma of the tongue removed 5.2 hours after the injection of BrdUrd.

on the time interval between labelling and biopsy. The advantage of v is that it obviates the need to correct LI and to make assumptions in the calculation of RM and λ. It should be noted that the absolute values of T_{pot} are different using the original Begg method and the v function; the latter giving longer T_S values and thus longer T_{pots}. However, the ranking of patients is similar with both methods.

In practice, a profile such as that shown in *Figure 5* can be obtained. The patient presented with a squamous cell tumour of the tongue that was biopsied 5.2 hours after an injection of BrdUrd. The tumour was aneuploid with a DNA index of 1.55; it can clearly be seen that virtually all the proliferation was associated with the aneuploid cells. This suggests that the first peak is pre-dominantly made up of normal stromal, vascular, or infiltrating cells. In some tumours this is not always the case, and the diploid component may also be proliferating indicating the presence of two tumour stemlines. In a specimen such as this the LI and T_S can be calculated for the aneuploid cells only. This will not be possible for a diploid tumour containing a mixture of normal and tumour cells, which the FCM is unable to distinguish from each other unless a cytokeratin marker is used with a third colour (this must be done on intact cells and not on nuclei preparations). In *Figure 5*, the redistribution of the labelled cells through the cycle is clearly demonstrated and the T_S was computed to be 12.5 hours for this tumour. The LI was 13.8%, and from these values a T_{pot} of 3.0 days was calculated assuming a λ value of 0.8 (this has been assumed from our studies of experimental tumours).

These measurements are being assessed in multi-centre clinical trials of radiotherapy to assess whether pre-treatment, cell-kinetic measurements can predict those patients whose tumours will do badly with the protracted conven-tional treatment schedules and who should be considered for more rapid (accelerated) regimes.

9 Conclusions

The development of antibodies to the halogenated pyrimidines combined with the attributes of FCM have led to a revolution in the study of cell kinetics and a resurgence of interest in their clinical application. The techniques are reproducible and relatively straightforward and can be tailored to individual experimental requirements. The early expectation that knowledge of tumour cell kinetics could improve the treatment of cancer is at last being realized.

Acknowledgements

This work is supported in its entirety by the Cancer Research Campaign. All the work depends on the expert skills of Christine Martindale, and the clinical studies would not be possible without the enthusiasm of Professor Stanley Dische and Dr Michele Saunders.

References

1. Steel, G. G. (1977). *Growth kinetics of tumours.* Clarendon Press, Oxford.
2. Murray, A. W. (1992). *Nature*, **359**, 599.
3. Pardee, A. B. (1989). *Science*, **246**, 603.
4. Wilson, G. D., McNally, N. J., Dische, S., Saunders, M. I., Des Rochers, C., Lewis A. A., and Bennett, M. H. (1988). *Br. J. Cancer*, **58**, 423.
5. Gratzner, H. G. (1982). *Science*, **218**, 474.
6. Dolbeare, F. A., Gratzner, H. G., Pallavicini, M. G., and Gray, J. W. (1983). *Proc. Soc. Natl Acad. Sci. USA*, **80**, 5573.
7. Wilson, G. D., McNally, N. J., Dunphy, E., Karcher, H., and Pfragner, R. (1985). *Cytometry*, **6**, 641.
8. Steel, G. G. and Bensted, J. P. M. (1965). *Eur. J. Cancer*, **1**, 275.
9. Schutte, B., Reynders, M. M. J., van Assche, C., Hupperts, P. S. G., Bosman, F. T., and Blijham, G. H. (1987). *Cytometry*, **8**, 372.
10. Beisker, W., Dolbeare, F. A., and Gray, J. W. (1987). *Cytometry*, **8**, 235.
11. Dolbeare, F. A. and Gray, J. W. (1988). *Cytometry*, **9**, 631.
12. Bakker, P. J. M., Stap, J., Tukker, C. J., van Oven, C. H., Veenhof, C. H. N., and Aten, J. (1991). *Cytometry*, **12**, 366.
13. McNally, N. J. and Wilson, G. D. (1986). *Br. J. Radiol.*, **59**, 1015.
14. Begg, A. C., McNally, N. J., Shrieve, D. C., and Karcher, H. (1985). *Cytometry*, **6**, 620.
15. Wilson, G. D. (1991). *Acta Oncologica*, **30**, 903.
16. White, R. A., Terry, N. H. A., Meistrich, M. L., and Calkins, D. P. (1990). *Cytometry*, **11**, 314.

Chapter 11

Analysis of cell proliferation using the bromodeoxyuridine/Hoechst–ethidium bromide method

Michael G. Ormerod* and Martin Poot[†]

*34, Wray Park Rd, Reigate, Surrey RH2 0DE, UK.

[†] University of Washington, Department of Pathology, Box 357705, Seattle, WA98195, USA

1 Introduction

The fluorescence of *bis*-benzimidazole dyes (Hoechst 33258 and Hoechst 33342) bound to DNA is quenched by the thymidine analogue, 5-bromodeoxyuridine (BrdUrd). Consequently, continuous labelling with BrdUrd and subsequent staining of DNA with Hoechst 33258 allows the discrimination of chromatids according to the number of replications they underwent during the observation period. The potential of this technique for analysing the progression of cells through their cycle has been established in several laboratories (see, for example, refs 1 and 2 and references therein). Cell-cycle kinetics are analysed using continuous BrdUrd labelling and bivariate Hoechst 33258/ethidium bromide (EB) flow cytometry. By the quenching of fluorescence from the Hoechst dye, cells are separated according to how many phases of DNA replication they have traversed, whereas the fluorescence of the non-quenched EB resolves the cell cycle into the G_1, S, and G_2 compartment. In this chapter, we refer to Hoechst/EB but propidium iodide (PI) can also be used in place of EB. Both dyes have been used to prepare the data shown in the figures.

Many studies have reported the analysis of the exit rate from the G_0/G_1 compartment of the cell cycle using cells that have been synchronized at this stage. The method is not limited to this type of analysis, however, and it has also been used to analyse cell-cycle progression in asynchronous cultures (see ref. 3 and references therein). Both applications are described in this chapter.

2 Cell culture and BrdUrd labelling

The method has been applied to a variety of cell types, each requiring specific, optimal cell-culture conditions. The concentration of BrdUrd added to the

179

culture must be optimized in order to label the DNA satisfactorily without causing cytotoxicity. Concentrations between 10 μM (human, ovarian carcinoma cell line) and 100 μM (human, peripheral blood lymphocytes) have been used. A preliminary experiment should be carried out in which cells are cultured for either 48 and 72 h (stimulated, quiescent cells) or 8 and 16 h (asynchronous cell cultures). After harvesting, the cells should be counted. Half the cultures should be stained for DNA using PI (see Chapter 6) to check for any adverse effect of BrdUrd on the cell cycle, and the other half stained with Hoechst 33258/EB (see below) to discover the lowest concentration which gives sufficient quenching of the Hoechst fluorescence. If cells accumulate in the G_2 phase of the cell cycle or there is a severe BrdUrd-dependent reduction in cell growth, lowering the BrdUrd concentration and/or adding an equimolar concentration of deoxycytidine can be tried.

The plating density should not exceed 3×10^3 cells/cm^2 for adherent cells or 2×10^5 cells/ml for suspension cultures. If these limits are exceeded, the BrdUrd may be depleted before the completion of the experiment. The possible movement of cells into the plateau growth phase during the course of the experiment must also be considered.

As BrdUrd is a strong photo- and radiosensitizer, cell cultures should be fully protected from short wavelength light by wrapping the culture flasks in aluminium foil. After the cells have been exposed to BrdUrd, all subsequent procedures should be carried out in dimmed light or under red illumination. Failure to prevent exposure of cells to light can lead to cell death and arrest of cells in the S and G_2 phases of the cell cycle.

After an appropriate period of culture, the cells are harvested and either resuspended in ice-cold staining buffer for analysis within a few hours, or resuspended in culture medium supplemented with 10% FBS and 10% dimethylsulfoxide and stored at $-20\,°C$ in the dark.

3 Cell staining and flow cytometry

Protocol 1

Labelling cells for BrdUrd/Hoechst/EB analysis

Reagents
Note: store all solutions in dark glass bottles at 4°C.

- Staining buffer: 100 mM Tris buffer, pH 7.4, 154 mM NaCl, 1 mM CaCl$_2$, 0.5 mM MgCl$_2$, 0.1% (v/v) Nonidet P-40 (BDH), 0.2% (w/v) BSA, 1.2 μg/ml Hoechst 33258. Prepare weekly from a 10 × concentrated stock solution in deionized water.

- Either EB or PI solution at 200 μg/ml in distilled water.

Method

1. Centrifuge a suspension of cells and resuspend the pellet in the staining buffer at between 4×10^5 and 10^6 cells/ml.

2. Incubate the cells for 15 min on ice in the dark.

3. Add EB or PI solution to give a final concentration of 2.0 µg/ml.

4. Incubate the cells for a further 15 min in the dark. Analyse the samples.

Stained nuclei do not show significant deterioration if kept on ice in the dark for up to 8 h; in some cases, the coefficient of variation (CV) of the peaks improves on storage. With some cells, the suspensions can be stored overnight.

The samples should be kept on ice and analysed between 0 and 4°C. If the flow cytometer does not allow the sample to be kept cool, it should be brought to room temperature before analysis since warming a cold sample may cause some drift during the initial minutes of analysis.

Flow cytometry can be performed on an arc lamp-based, epi-illumination system (e.g. Partec PAS II or III) or on any type of flow cytometer equipped with a laser emitting in the UV. On the arc lamp-based instruments appropriate excitation light is selected with a UG 1 and a BG 38 filter (Schott) and an FT 450 dichroic mirror. Hoechst fluorescence (blue) is selected with a K 45 filter and an FT 510 dichroic mirror, and EB or PI fluorescence (red) is collected with a K 65 LP filter. On laser-based instruments, blue and red fluorescence are collected using the appropriate filters, typically using an LP dichroic to remove any scattered UV, an LP 510 dichroic mirror to separate blue and red light which are further selected with 460 nm bandpass and 600 nm long-pass filters, respectively.

A bivariate cytogram of red vs. blue fluorescence should be recorded. To obtain the best resolution with synchronized cells, the gain on the blue channel should be adjusted to position cells in G_0/G_1 about 80% along the axis (for example, channel 800 in a 1024-channel display) and that of the red to about 45% (channel 460 in 1024). With asynchronous cells, both channels should be adjusted to about 45%.

An optimal CV is generally obtained at flow rates between 200 to 500 nuclei/ sec.

In the arc lamp-based instruments and in instruments using an argon-ion laser as a source of UV, EB (or PI) is excited via resonance energy transfer from the Hoechst 33258 (see Chapter 2, Section 2). This effect results in some quenching of the red fluorescence in the presence of BrdUrd. He–Cd lasers have an output at 325 nm, which excites EB and PI directly, giving better resolution of the cell cycle (4).

4 Synchronized cells

4.1 Cell cultures

These experiments generally use either naturally quiescent lymphocytes from peripheral blood, spleen, or thymus or cultured cells rendered quiescent by the reduction of growth factors.

Lymphocytes need only to be isolated by density centrifugation on Ficoll–Hypaque (see, for example, Chapter 3, Section 2). Anchorage-dependent cells are rendered quiescent by allowing them to reach contact inhibition of growth and subsequent exposure to a low serum concentration. The exact conditions depend on the cell type. Diploid human fibroblasts reach quiescence after 48 h in 0.1% FBS, but NIH-3T3 cells tolerate only 24 h at 0.5% FBS. Epstein–Barr virus-transformed, B-lymphoblastoid cells reach quiescence if they are left without replenishment of the culture medium for 4 days or longer.

Typically, lymphocytes are aliquoted into tubes at 5×10^4 cells/ml in RPMI-1640 culture medium supplemented with 16% FBS, 200 μg/ml phytohaemagglutinin, 100 μM BrdUrd, and 100 μM deoxycytidine. The tubes are placed upright in the incubator at 37°C, which allows maximum cell-to-cell contact thus improving cell proliferation. Adherent cells are trypsinized and replated at a density of 10^3–3×10^3 cells/cm^2 of culture flask in culture medium supplemented with the appropriate concentrations of FBS, BrdUrd, and deoxycytidine.

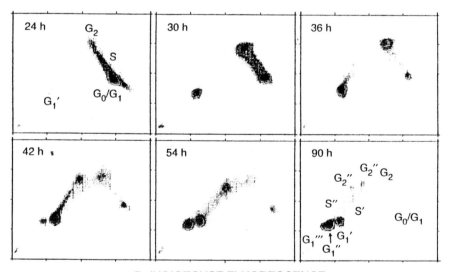

BrdU/HOECHST FLUORESCENCE

Figure 1 Bivariate cytogram of human diploid fibroblasts after continuous BrdUrd labelling for up to 90 h. The abscissa is BrdUrd-quenched Hoechst fluorescence (blue) and the ordinate is unquenched EB fluorescence (red). From right to left the arrays of clusters represent the first (G_0/G_1, S, G_2), second (G_1', S', G_2'), and third cycle (G_1'', S'', G_2'') after serum stimulation of quiescent cells. The lowermost clusters are the G_1 phases, and the uppermost are the G_2 phases of each cell cycle. (Data recorded using a PHYWE ICP-22 cytometer with a mercury arc lamp.)

The drug is usually added at the start of the experiment if a cytotoxic agent is being studied.

4.2 Quantitative analysis of data

Figure 1 displays a typical series of bivariate cytograms obtained using human fibroblasts. At 12 h after serum stimulation, the cells appeared in a single cluster which represented quiescent or G_0/G_1 cells (not shown). At 24 h after serum stimulation, the signal track moving from the G_0/G_1 cluster to the left represented cells in the first S phase. The G_1' cluster is made up of cells which traversed a full cell cycle and underwent mitosis. Cell division caused halving of the fluorescence intensity in both the Hoechst and EB direction. The signal track of cells in the second S phase moved to the right because of bifiliary substitution of the DNA with BrdUrd (see panels from 36 h on).

The bivariate cytogram resolves cells into three successive cell cycles and additionally separates the G_1, S, and G_2 compartments of each cell cycle. Each component cell cycle can be isolated from the cytogram by setting a region. Displaying a histogram of EB (or PI) fluorescence gated on this region yields a conventional cell-cycle distribution, which can be analysed with an algorithm of the kind discussed in Chapter 6. Thus, the distribution of cells among cell-cycle compartments at a given time point can be calculated.

5 Asynchronous cells

5.1 Cell culture

BrdUrd is added to cultures (suspension or adherent) of cells which have been established in the early stages of exponential growth. If the effects of radiation (high energy, UV, or heat) are to be studied, the cultures should be treated before adding BrdUrd. Short-term drug treatments should also be carried out before adding BrdUrd; for continuous treatment with a cytotoxic drug, the compound should be added with the BrdUrd.

5.2 Quantitative analysis

Figure 2 shows data obtained from an exponentially growing culture of a human embryonic fibroblast cell line (MRC5/34) incubated with BrdUrd for 0, 8, 16, and 32 h. Initially both red (PI) and blue (Hoechst) fluorescence gave a normal cell cycle with cells in G_1, S, and G_2/M. After 8 h, cells in G_2 had reduced due to cell division. Cells in S phase at time 0 had taken up BrdUrd so that, while their red fluorescence increased in line with DNA content, their quenched blue fluorescence did not increase. Cells in late S had moved into G_2 (labelled G_2^*); cells moving from G_1 into S have been labelled Sf. After 16h, cells from early and mid S had reached G_2^*; cells from G_1 which had moved into G_2 are labelled G_2'. At 32 h, many of the cells originally in G_1 had cycled once and returned to G_1 (G_1^*).

Numerical data can be obtained by drawing regions around the different cell-cycle compartments (3). Because of the overlap between the G_1 and early S

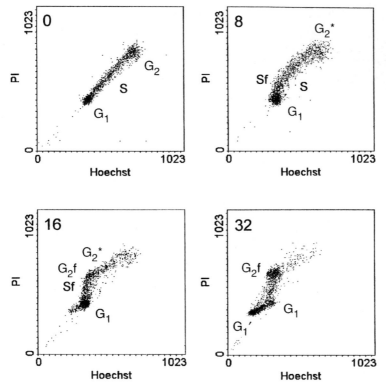

Figure 2 BrdUrd–Hoechst/PI cytograms obtained from an asynchronous culture of a human embryonic fibroblast cell line (MRC5/34). 40 μM BrdUrd plus 40 μM deoxycytidine was added at time 0. The numbers represent the time in h. (Cells were prepared by Dr D. Gilligan and data recorded by Mrs J. C. Titley on a Coulter Elite ESP using an argon-ion laser.)

phases and G_2 and late phases, this method will overestimate the G phases and underestimate S phases. Unfortunately, it is difficult to avoid this source of error without recourse to complex computer programs. The data should be corrected to take into account that, as cycling cells divide, they dilute any undivided cells in the culture.

Figure 3 shows examples of blocks in the cell cycle. The power of the method lies in the ability to determine from which phase of the cell cycle the blocked cells originated. Panel B shows a G_2 block 28 h after irradiating MRC5/34 cells with 5 gray of γ radiation. Cells which were irradiated in G_2, S, and G_1 phases of the cell cycle are all blocked in G_2 (labelled G_2, G_2^*, and G2f, respectively). Compare this result to that shown in panels C and D. The data is from a human lymphoblastoid cell line (W1L2) incubated with BrdUrd for 24 h. The cells in panel D had been incubated for 2 h with 4 μM of the chemotherapeutic drug, cisplatin, prior to the addition of BrdUrd. Although there is a G_2 block, the blocked cells are all in G_2f — they were cells that had been incubated with the drug in G_1 of the cell cycle. Cells incubated with the drug in S and G_2 phases had divided normally.

184

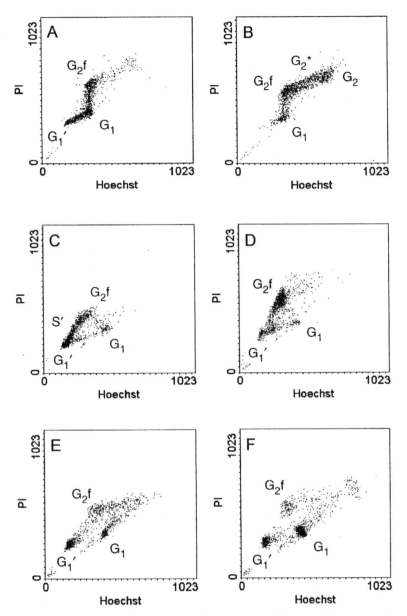

Figure 3 BrdUrd–Hoechst/PI cytograms demonstrating blocks in the cell cycle. Panels A and B: a human embryonic fibroblast cell line (MRC5/34) 28 h after the addition of BrdUrd; (A) no irradiation; (B) 5 gray γ radiation. (Cells were prepared by Dr D. Gilligan.) Panels C and D: human lymphoblastoid cell line (W1L2) 24 h after the addition of BrdUrd; (C) no treatment; (D) incubated at time 0 for 2 h with 4 μM cisplatin. Panels E and F: human cervical carcinoma cell line (283) 26 h after the addition of BrdUrd; (E) no irradiation; (F) 24 h after 4 gray γ radiation. (Cells were prepared by Mrs C. Bush.) (Data recorded by Mrs J. C. Titley on a Coulter Elite ESP using an argon-ion laser.)

185

Panels E and F show data from a human cervical carcinoma cell line (283) that underwent a G_1 block after γ radiation. After 26 h, 40% of the irradiated cells (panel F) have not left G_1 compared to 11% in the unirradiated control (panel E). Only a proportion of the cells irradiated in G_1 became blocked in G_1, some of the G_1 cells progressed through the cycle and divided (in compartment G1'). Presumably these were cells irradiated in late G_1 phase.

6 Troubleshooting

Typical results as shown in *Figures 1* and *2* might not be obtained due to several types of pitfalls:

(a) *Cytotoxicity of BrdUrd*. Continuous exposure to a halogenated nucleoside may entail a plethora of adverse effects on cellular functions, including metabolic changes due to alterations in the balance of nucleotide pools, direct DNA damage, and alterations in DNA–protein interaction. Directly or indirectly, these effects could impair proliferation. It is recommended that each cell type should be tested with a series of BrdUrd concentrations before using the method (see also (e) below).

(b) *Distorted cytograms*. Initial tuning of the instrument is best performed with unstimulated lymphocytes or with cells without BrdUrd from the cell culture under study. If focusing of G_0/G_1 cells to a narrow, round cluster is difficult, the sample delivery channel of the flow cytometer might be dirty. This can occur after many samples stained with EB have been analysed, as EB tends to lyse some nuclei, which subsequently stick to the tubing.

(c) *Clumping of nuclei*. If the nuclei are clumped, pass the samples through a 30 μm gauze either before or after staining. Treatment of samples with nucleases or proteases is not recommended, as these enzymes will continue to act, and thus to degrade nuclei, in the staining buffer.

(d) *Poor quenching of Hoechst fluorescence*. This might either be due to insufficient BrdUrd incorporation into DNA or to inappropriate staining. The first phenomenon can occur if the serum batch in use contains significant amounts of thymidine, which will compete with BrdUrd for the nucleotide salvage pathway. The remedy is to increase the BrdUrd concentration in the culture medium, while controlling for possible cytotoxicity due to excess BrdUrd (see above). If increasing the BrdUrd level does not improve the cytogram, a very low A–T content of the DNA might be the cause of the trouble. To control this factor, use cells with a known A–T content (e.g. human lymphocytes or mouse 3T3 cells), which should give the kind of quenched fluorescence from Hoechst shown in the figures. If, with human or mouse cells, a bad quenching of Hoechst fluorescence is obtained, the staining procedure might be inappropriate. Increasing the concentration of EB or PI and/or lowering the Hoechst concentration might improve the cytogram.

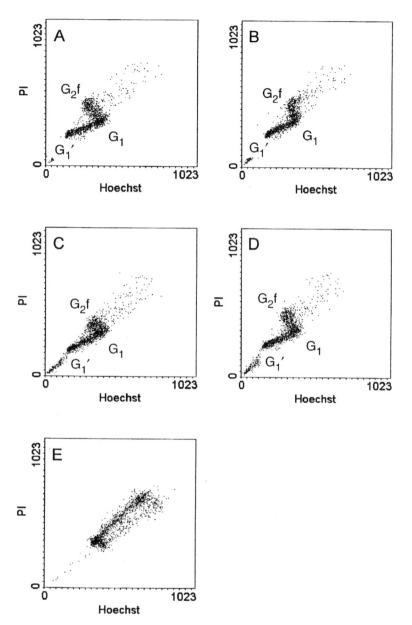

Figure 4 BrdUrd–Hoechst/PI cytograms demonstrating various problems. Panels A, B, C, and D: human cervical carcinoma cell line 15 h after the addition of BrdUrd; (A) 60 μM BrdUrd, 10^7 cells/ml; (B) 20 μM BrdUrd, 10^6 cells/ml; (C) 30 μM BrdUrd, 10^7 cells/ml; (D) 30 μM BrdUrd, 10^6 cells/ml; (cells prepared by Mrs C. Bush). (E) human embryonic fibroblast cell line, no BrdUrd; demonstrating effect of incomplete lysis; (cells prepared by Dr D. Gilligan). (Data recorded by Mrs J. C. Titley on a Coulter Elite ESP using an argon-ion laser.)

(e) *Incorrect concentration of BrdUrd*. Panels A and B in *Figure 4* show the effect of two different concentrations of BrdUrd. For asynchronous cells, the lower concentration of BrdUrd to give a cytogram of the appearance shown in panel B is preferable.

(f) *Too high a cell concentration*. If the cell concentration is too high, there may be insufficient PI (or EB) in the sample to give an acceptable cell cycle. This effect is shown in *Figure 4*, panel C. After diluting the cells in the staining buffer, the resolution of the cell cycle improved (panel D).

(g) *Incomplete lysis of the cells*. If the cells are not all lysed, a 'shadow' may appear in the cytogram (see *Figure 4*, panel E). A brief incubation (1–3 min) at 37 °C will usually resolve the problem.

Acknowledgements

MP acknowledges the help of Miss J. Köhler with the preparation of *Figure 1*. MGO thanks his colleagues, Mrs J. C. Titley, Dr D. Gilligan, and Mrs C. Bush, all at the Institute of Cancer Research, Sutton, UK, for permission to use their data to prepare *Figures 2–4*.

References

1. Rabinovitch, P. S., Kubbies, M., Chen, Y. C., Schindler, D., and Hoehn, H., (1988). *Exp. Cell Res.*, **74**, 309.
2. Poot, M., Kubbies, M., Hoehn, H., Grossman, A., Chen, Y., and Rabinovitch, P. S. (1990). In *Methods in cell biology*, Vol. 33 (ed. Z. Darzynkiewicz and H. A. Crissman), p. 186. Academic Press, New York.
3. Ormerod, M. G. and Kubbies, M. (1992). *Cytometry*, **13**, 678.
4. Kubbies, M., Goller, B., and Van Bockstaele, D. R. (1992). *Cytometry*, **13**, 782.

Chapter 12

Chromosome analysis and sorting by flow cytometry

Derek C. Davies,* Simon P. Monard,† and
Bryan D. Young†

*Flow Cytometry Laboratory, ICRF, Lincoln's Inn Fields, London WCA 3PX, UK
†Trudeau Institute, Algonquin Avenue, Saranac Lake, NY 12983, USA
†ICRF Medical Oncology Unit, St Bartholomew's Hospital, Charterhouse Square, London EC1A 7BE

1 Introduction

Many types of cancer and genetic disease are characterized by chromosomal abnormalities. These can be detected by conventional cytogenetics, which involves photographing banded chromosomes on metaphase spreads. Although this is a widely used technique in haematology, oncology, and pre-natal diagnosis, it is a time-consuming process that relies heavily on the skill and experience of the cytogeneticist. In addition, in cancerous cells, the sometimes complex karyotypes encountered mean that such analysis is often extremely difficult. An alternative approach is to prepare a monodispersed suspension of chromosomes, stain with one or two fluorescent DNA dyes, and pass them through a flow cytometer. The intensity of the fluorescence signal from each chromosome is recorded, the values being dependent on their DNA content. In single-colour analysis, a non-base-specific dye such as ethidium bromide is used. Accumulating data from 50 000 chromosomes and presenting this as a histogram of fluorescence intensity produces a distinctive species-specific pattern of peaks. However, not all chromosomes appear as a single peak. An improvement is to use bivariate analysis that exploits the base-pair binding preferences of DNA specific dyes: usually Hoechst 33258 (which has an adenine/thymidine (AT) binding preference) in combination with chromomycin A_3 (which has a guanine/cytosine (GC) binding preference). The intensity of staining with these fluorochromes is dependent not only on the DNA content but also on the base composition of each chromosome. The data accumulated can be presented as an isometric plot or, better, as a dot plot or contour map (see *Figure 1*). The bivariate plot resolves all the human chromosomes as separate peaks, with the exception of chromosomes 9, 10, 11, and 12.

The first human flow karyotypes were demonstrated using chromosomes isolated from fibroblast cells (1, 2). Subsequently, high-resolution flow karyotypes

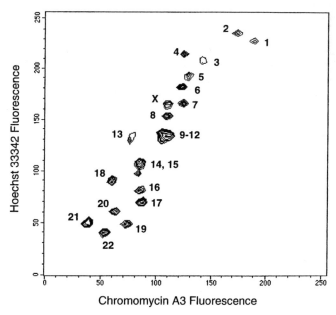

Figure 1 Bivariate flow karyogram of a cell line derived from a normal individual. All the chromosome types can be resolved with the exception of the 9–12 group. This example is from a human female.

were obtained from short-term, phytohaemagglutinin (PHA)-stimulated, peripheral blood lymphocytes (3) and lymphoblastoid cell lines (4). In practice, flow karyotypes can be obtained from most types of cells whether grown in suspension or monolayers.

Since the introduction of flow cytogenetics, the technology has been refined and developed and has provided an essential resource for the analysis of the human genome. Although a karyotype using a bivariate system provides no information about the individual cells, it can provide an accurate measurement of the frequency of chromosome types. For example, trisomy 21 would be seen as a 50% increase in the number of events in the chromosome 21 peak. The presence of an abnormal chromosome may also be determined, provided there is a difference in either DNA content or AT:GC ratio between the abnormal chromosome and its normal counterpart. In addition, translocations resulting in two derivative chromosomes, which differ in either DNA content or base-pair ratio from the chromosomes from which they are derived, will appear as two separate peaks in positions where there are usually none (see *Figure 12.2*). Small deletions and/or marker chromosomes can also be detected. With these advantages, why has flow cytogenetics based on a rapid, highly reproducible, machine-based approach not replaced conventional cytogenetics? There are several reasons: sometimes the derivative chromosomes can appear in the same position as other normal chromosomes making them difficult to detect; also a reciprocal translocation will remain undetected if it results in derivative chromosomes of the same size and base-pair ratio as the originals. The main

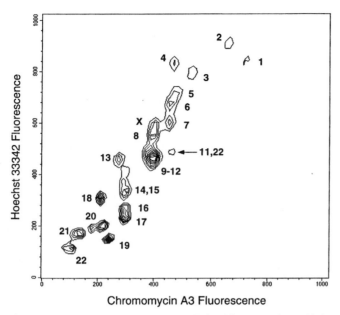

Figure 2 Flow karyogram of a cell line derived from a patient with lymphoma. Note that there is a marker chromosome. Flow sorting markers such as this followed by reverse chromosome painting can be used to rapidly identify their origin. In this case the derivative is from an 11, 22 translocation.

reason is that flow cytogenetics is greatly complicated by the high level of polymorphism in the general population. Certain chromosomes have regions of centric heterochromatin, which can vary considerably in size, resulting in microscopically visible differences (5, 6). So, in practice, it is difficult to ascertain whether differences are normal or due to translocations or deletions. However, recent developments in the use of flow cytometry for the karyotyping of individual cells (7) may well represent a significant technological advance.

The real strength of flow cytometry is its ability to physically separate—'sort' —the different chromosome types with a high degree of purity. Any chromosome that can be resolved as a separate peak can be sorted with a high degree of purity—over 98% should be possible. Flow-sorted chromosomes have proved very useful in the production of phage libraries (4) and, more recently, cosmid libraries (8). The construction of gene libraries requires relatively large numbers of chromosomes, $2-5 \times 10^6$ typically being required, but these libraries have proved to be an essential resource in the study of the human genome. As large numbers of chromosomes are required, a large number of starting cells is needed and suspension cell lines are the most convenient in this case.

The development of the polymerase chain reaction (PCR) provides the opportunity of performing molecular studies on small numbers of chromosomes. The use of primers directed against the human Alu-repeat sequence used with flow-sorted chromosomes as the template DNA thereby generates an array of products representing most of the chromosome (the centromere and the Giemsa dark-stained bands of chromosomes are relatively poorly represented). These PCR products can be used in two main ways. First, they can be cloned into a plasmid vector and used to probe cosmid libraries (9, 10) and second, small numbers of flow-sorted chromosomes can also be used to generate chromosome-specific 'paints' after degenerate oligonucleotide-primed PCR (DOP-PCR) (11) or Alu-mediated PCR (ref. 12; see *Section 5*). In this approach the library (containing amplified DNA fragments ranging in size from 500 bp to 3 kbp) is labelled with fluorescent dUTPs or biotin–dUTPs and used to perform *in-situ* hybridization on metaphase spreads. The resultant fluorescence signal will be restricted to the original sorted chromosome. Labelled PCR products from normal chromosomes can be applied to abnormal metaphase spreads to identify any translocations present that may otherwise go undetected. This is known as forward chromosome painting. A rapid way of identifying an abnormal marker chromosome is to flow-sort it, generate paints, and subsequently perform *in-situ* hybridization on metaphase spreads from normal individuals. The resultant signal will be restricted to those chromosomes from which the marker chromosome is derived. This approach has been termed reverse chromosome painting.

The presence or absence of a particular DNA sequence in flow-sorted chromosomes can also be assessed. For instance, the sequence across the breakpoint of a marker chromosome, if it is known, can also be assessed by PCR with primers specific to that sequence. Such is the power of the PCR technique that it is possible to generate a detectable product using a single flow-sorted chromosome as template DNA. In such experiments, our experience suggests that sorting directly into the PCR buffer gives the most consistent amplifications.

2. Preparation of chromosome suspensions

The key to making a good chromosome preparation is to start with a cell culture of healthy cells growing exponentially. Although several preparative techniques have been developed (13, 14), they all have certain common features. A culture of growing cells is treated with an agent such as colcemid or vinblastine in order to arrest sufficient cells in metaphase. The period of treatment depends on the cell-cycle time, so it is important to determine the optimal time. Extensive culture in the presence of such agents can lead, first, to a high degree of chromosome contraction and ultimately to cell death and necrosis resulting in increased debris in the preparation. If the cells grow as an attached layer, mitotic shake-off can be used to obtain an enriched population of metaphase cells. However, it is important that the cells are not confluent or near confluent before blocking as insufficient cells will be available for the release of chromosomes.

Sufficient chromosomes for flow-karyotyping can also be obtained from as little as 2 ml of human peripheral blood. The use of such short-term cultures has the advantage that there is little opportunity for the karyotype alterations that can occur in established cell lines (15, 16). Also, in this case, the flow karyotype can be compared directly with the conventionally banded karyotype normally performed for routine analysis. Lymphoblastoid cell lines, on the other hand, represent an ideal source when large numbers of chromosomes are required, such as for sorting for library construction. Some carry known aberrant chromosomes that may be sorted to facilitate analysis of the genotype of the aberration or to simplify subchromosome gene mapping with known karyotype abnormalities. It should be remembered, however, that transformed cell lines, especially rodent cell lines, often develop chromosomal aberrations not present in the primary tissue from which they were derived.

There are two main methods used for preparing chromosomes and the choice of method depends on the aims of the particular experiment. The hexylene glycol method (13) is reported to allow banding and hence identification of the chromosomes following sorting, but the discrimination of the different chromosome types on a commercially available flow cytometer is relatively poor. The preparative technique used in our laboratory is described in *Protocol 1*. This method, which uses polyamines to stabilize the chromosomal DNA (17), has the particular advantage that very high molecular weight DNA can be obtained after sorting (14). The discrimination of the different chromosomes is good, making the method ideal to use when sorting chromosomes for cosmid library construction and for sorting prior to PCR amplification. *Protocol 2* (18) is thought to give the best discrimination between chromosome types. Although, as stated above, chromosomes can be prepared from almost any culture of growing cells, we find newly established lymphoblastoid cell lines the most convenient to work with.

Protocol 1

Chromosome preparation: polyamine method

This is the method used in our laboratory; the preparations can be stored and used for several weeks or even months, the DNA is of high quality, the resolution between chromosome types is good, and the same protocol can be used for preparing chromosomes from human or rodent cells.

Equipment and reagents

- 50 ml conical tubes
- Fluorescence microscope
- Colcemid (Life Technologies)
- Hypotonic solution: 75 mM KCl
- 10 × chromosome isolation buffer (CIB): 20 mM NaCl, 80 mM KCl, 15 mM Tris–HCl, 0.5 mM EGTA, 2 mM EDTA, 0.15% (w/v) 2-mercaptoethanol, 0.2 mM spermine (free base), 0.5 mM spermidine (free base) pH 7.2 in autoclaved distilled water
- CIB1: 1.2 mg/ml digitonin (Sigma) in CIB (see step 6)

- 0.22 μm sterile disposable filter
- 12 × 75 mm plastic tubes
- 50 μg/ml propidium iodide (Sigma) in phosphate buffered saline (PBS), pH 7.2
- 50 μg/ml propidium iodide (Sigma) in PBS, 0.1% (w/v) Triton-X 100
- 100 μg/ml Hoechst 33258 in distilled water
- 15 mM MgCl$_2$
- 2 mg/ml chromomycin A$_3$ in ethanol

Protocol 1 continued

Method

1. Block cells[a] with 0.05 µg/ml colcemid for 5–16 hours depending on the rate of growth (usually blocking overnight gives good results).

2. Estimate the proportion of suspension cells in mitosis by pelleting the cells (150g, 5 min room temperature) from 1 ml of the blocked cell culture, discarding the supernatant, and resuspending in PBS containing 50 µg/ml propidium iodide (PI) and 0.1% Triton X-100. Use either a fluorescence microscope or a bench-top flow cytometer for this estimation. It should be possible to get 40–60% of the cells in mitosis with suspension cell lines.

3. Estimate the proportion of monolayer cells in mitosis on an inverted microscope, mitotic cells are round and can usually be shaken off into the medium by giving the flask a sharp rap. Some monolayer cell lines may require the use of trypsin.

4. Once in suspension, centrifuge the cells at 100 g for 10 min at room temperature in 50 ml plastic tubes, discard the supernatant, and resuspend the cells in fresh medium and centrifuge again for 10 min at 100 g.

5. Discard the supernatant by inverting the tube. Remove the last few drops from inside the tube with a tissue. Disaggregate the cell pellet by vortexing gently or by flicking the tube. Add 5 ml of hypotonic solution, mix gently, and leave for 10–30 min at room temperature (lymphoblastoid cell lines generally require less time than fibroblastoid cell lines). This is a convenient time to pool the contents of several tubes. Centrifuge the tubes for 10 min at 100 g at room temperature.

6. Meanwhile, prepare chromosome isolation buffer 1 (CIB1) by adding 1 ml of 10 × CIB, 9 ml of distilled water, and 12 mg of digitonin together with a small magnetic 'flea' to a small beaker, warm gently on the hotplate stirrer until the digitonin has dissolved. Because some batches of digitonin seem to be more difficult to dissolve and rather than risk denaturing some components of the CIB, dissolve the digitonin in 5 ml water at approx. 60 °C, allow it to cool, then add the 10 × CIB and make the volume up to 10 ml. Adjust the pH to 7.2 using 0.1 M NaOH. Filter through a 0.22 µm sterile disposable filter into a new tube and place on ice. Make the chromosome suspensions very concentrated, i.e. no more than 10 ml of CIB1 for 200 ml of a suspension cell culture, to enable a high sample rate to be used.

7. Carefully remove the supernatant with a Pasteur pipette and flick the tube gently to disaggregate the cells. Add 10 times the volume of the cell pellet in cold CIB1, flick the tube gently. Mix a small amount of the preparation with an equal volume of PBS with 50 µg/ml PI and view with a fluorescence microscope. If the chromosomes are not monodispersed, flick the preparation more vigorously or aspirate with a plastic Pasteur pipette. Use gentle vortexing if required, but, if possible, avoid vigorous vortexing as it can result in chromosome breakage. The chromosome suspension may be stored at 4 °C for several weeks with little deterioration of flow karyotype.

8. Transfer 1 ml of the chromosome suspension into a 12 × 75 mm plastic test-tube,

Protocol 1 continued

add 30 μl Hoechst 33258 (100 μg/ml in distilled water), mix immediately. Add 40 μl of 15 mM $MgCl_2$ and 50 μl chromomycin A_3 (2 mg/ml in ethanol), mix and leave the sample at 4°C for 2 hours in the dark.

[a] Cell lines, either monolayer or suspension may be used, as may phytohaemagglutinin-stimulated peripheral blood cells. Whichever type of cell is used, the best preparations will be made from healthy cells growing optimally. Subculture cells 24 hours before blocking with colcemid.

[b] The chromosome profile can be improved if 100 μl sodium citrate (100 mM) and 100 μl sodium sulfite (250 mM) are added at least 15 min prior to running on the cytometer. Aggregates in the sample can be removed by centrifuging at 200 g for 1 min, then transferring the supernatant to a new tube. Centrifugation selectively depletes the larger chromosomes so should be avoided if the flow karyotype is to be analysed.

Protocol 2

Chromosome preparation: magnesium sulfate method

Equipment and reagents

- Chromosome isolation buffer 2 (CIB2): 40 mM KCl, 5 mM Hepes, 10 mM $MgSO_4$, 3 mM dithiothreitol, pH 8 in autoclaved distilled water
- Triton X-100 (Sigma)

- 50 ml conical tubes
- 12 × 75 mm plastic tubes
- Colcemid (Life Technologies)
- Phase-contrast microscope

Method

1. Prepare colcemid-blocked cells as in *Protocol 1*.

2. Centrifuge the cells at 300 g for 10 min at room temperature, decant the supernatant, and drain the tubes on an absorbent paper towel.

3. Add 1 ml of CIB2 to 6×10^5 cells, resuspend gently, and incubate at room temperature for 10 min.

4. Add 0.1 ml of Triton X-100 solution (2.5% in distilled water) and incubate on ice for 10 min.

5. Vortex for 10–20 seconds to disrupt the cells (monitor using phase-contrast microscopy).

6. Stain for bivariate analysis as in *Protocol 1*, step 7.

The choice of stain for chromosome sorting and analysis depends largely on the configuration of the flow cytometer and, in particular, on the light source available. Bivariate analysis using Hoechst 33258 and chromomycin A_3 offers the clearest separation of the different chromosome types. If a single laser cytometer is to be used there are three main choices of DNA stain. The most commonly used are ethidium bromide or propidium iodide, which can be excited using a 488-nm light source and have no base preference. Second, one of the

UV-excited dyes such as Hoechst 33258 or DAPI can be used. These dyes have an AT-binding preference. Last, chromomycin or mithramycin can be used, these dyes have a GC-binding preference and can be excited by the 457.9-nm line of an argon laser.

Caution: many of the reagents used in the preparation and staining of chromosomes are highly toxic or potentially mutagenic. Care should be taken and disposable gloves should be used at all times.

3 Instrumentation

Univariate chromosome analysis using ethidium bromide or propidium iodide can be performed on most flow cytometers equipped with an argon-ion laser and having at least 256 channels available per measured parameter. The emission signal can be collected using a 580 nm long-pass filter. Other fluorochromes (e.g. DAPI, Hoechst 33258) can be used for univariate analysis and sorting if they give a clearer separation of the chromosome peak of interest and if an appropriate excitation wavelength is available.

Bivariate analysis is preferable as more of the different chromosome types are represented as separate peaks. Chromosomes are usually stained with Hoechst 33258 and chromomycin A_3 (18). Bivariate chromosome analysis and sorting requires a flow cytometer equipped with two argon-ion lasers. The primary laser, i.e. the laser that intersects the sample stream nearer the nozzle, is tuned to multi-line UV (generally around 350–370 nm) and is used to excite the Hoechst dye. The secondary laser is tuned to 457.9 nm and is used to excite the chromomycin. The signal from the Hoechst dye is collected using a 390 nm long-pass and a 480 nm short-pass filter. The chromomycin signal is collected using a 490 nm long-pass filter only. All filters are of the coloured glass variety. The signal from the primary laser is usually designated fluorescence 1 (FL1) and that from the secondary laser fluorescence 2 (FL2).

The alignment of the laser beams is of critical importance. Special care should be taken to ensure that they do not pass too close to the edge of any of the prisms as this can cause diffraction and subsequent loss of resolution. It is important that the lasers are functioning in the TEM00 mode. The laser focusing lens may be an achromatic doublet, i.e. two lenses of different materials positioned close to each other or cemented together. If the lens is of the cemented type, the cement may discolour after prolonged use in the UV and so should be inspected periodically. The usual nozzle orifice size used is 50 μm, although a 70 μm nozzle can be used. The cleanliness of the nozzle is of paramount importance—a dirty nozzle can cause poor resolution, increased noise, and irregular deflection streams. When sorting chromosomes, particularly for library construction, keeping the sample and sorted fraction cool with ice or circulating cold water reduces the chance of DNA degradation.

Bivariate chromosome analysis can be performed using only the two fluorescence parameters FL1 and FL2, but it can be useful to use one or more light-scatter parameters for gating out debris. The instrument should be triggered on the Hoechst signal and the minimum possible threshold should be used, as a high threshold would allow small particles such as chromosome fragments to pass undetected through the instrument and into the sorted droplets. It is advisable to do a test sort before commencing any chromosome sorting. This can be done on fluorescent microspheres or preferably on chromosomes. The sorted chromosomes can either be re-stained with Hoechst and chromomycin A_3 and re-run on the same instrument or if a bench-top analyser is available, the chromosomes can be stained with propidium iodide and analysed on this instrument. Bench-top cytometers, although unable to identify sorted chromosomes, are able to indicate if the sorted fraction is composed of a single chromosome type and hence whether the instrument is sorting efficiently. If the sorted sample is re-stained and run on the same instrument, care should be taken that the sample tubing is free from unsorted chromosomes; it may be necessary to replace the sample tubing.

Sorting chromosomes for library construction is a time-consuming process. With a good preparation and a well set-up instrument it should be possible to pass 1500–2500 chromosomes though the instrument per second without deterioration in resolution. Thus it should be possible to sort a single copy chromosome at a rate of 30–50 per second. When running chromosomes for analysis the best discrimination between chromosome peaks will be obtained using a low sample rate. However, this can still take hours, if not days, of continuous sorting. More appealing is the combination of PCR and flow sorting, which only takes a few minutes sorting.

If there is poor discrimination between chromosomes seen on the cytometer it is not always easy to determine whether this is due to a poor preparation or poor cytometer performance. The performance of the instrument can be monitored using fluorescent microspheres. The coefficients of variation (CVs) of the fluorescent peaks and the intensity of the signal give an indication of how the instrument is performing. Staining the chromosome preparation with propidium iodide and viewing with a fluorescence microscope will show the quality of the preparation, and whether there are too few chromosomes or if the chromosomes are aggregating. Finally, it is invaluable to compare the results obtained with those of another laboratory where chromosome sorting or analysis is performed routinely.

For sorting, use a sheath buffer containing 100 mM NaCl, 10 mM Tris–HCl, 1 mM EDTA, pH 8 in distilled water.

Caution: care should be exercised when aligning the lasers especially in the UV. The lowest laser powers should be used and protective goggles must be worn.

4 Flow sorting chromosomes for library construction

Cosmid libraries constructed from flow-sorted chromosomes (8) have proved to be a vital resource for the analysis of the human genome. Flow sorting for such a purpose can be time-consuming, so it is essential to sort at the highest rate possible without compromising purity. A concentrated chromosome suspension allows a higher sample rate for the same sample pressure. Keeping both the sample and sorted fractions cool also reduces the chance of DNA degradation.

Although the human chromosomes 9–12 cannot be separated using univariate or bivariate analysis, it is, however, possible to use hamster/human hybrid cell lines with a single human chromosome to sort such a chromosome from the hamster background (19). A hamster background is preferable to a mouse background as hamster chromosomes are fewer and therefore more clearly separated. Another approach to isolating chromosomes in the 9–12 peak is to select a cell line that carries a deletion on the chromosome of interest causing it to appear as a separate peak. Time can also be saved if a cell line with a duplication of the chromosome of interest is used.

Protocol 3

Preparation of chromosomes for library construction

Equipment and reagents
- Sheath buffer: 100 mM NaCl, 10 mM Tris–HCl, 1 mM EDTA, pH 8 in distilled water
- t-RNA (Life Technologies)
- Proteinase K (BDH)
- Stock solution of 500 mM EDTA pH 8
- Stock solution of 20% (w/v) *n*-lauroylsarcosine (sodium salt)
- Sterile 1.5 ml conical tubes with screw caps
- Working solution: 250 mM EDTA, 10% (w/v) *n*-lauroylsarcosine

Method
1. Prepare the sterile sheath buffer containing 500 μg/ml of t-RNA, dispense 50 μl of this solution into sterile 1.5 ml conical tubes and vortex vigorously to coat the inside of the tubes. If required, store the tubes by freezing them quickly on dry ice and then place in a freezer at –20°C.
2. Sort 5×10^5 chromosomes into each tube.
3. Add working solution to the sorted chromosome suspension to make a final concentration of 25 mM EDTA and 1% *n*-lauroylsarcosine. When using a 50 μm nozzle and a three-drop deflection, 5×10^5 chromosomes should occupy a volume of about 700–800 μl, thus add 70–80 μl working solution.
4. Add 180 μg of Proteinase K to each tube, vortex, and incubate at 42°C overnight.
5. If required, store the resulting DNA preparation at 4°C (stable for many months).

5 Generation of chromosome paints

5.1 Degenerate oligonucleotide-primed polymerase chain reaction (DOP-PCR)

It is possible to generate chromosome paints from flow-sorted chromosomes from any species by DOP-PCR using the primer 6-MW (11). The six specific bases at the 3′ end of the oligonucleotide theoretically prime every 4 kbp along the template DNA at the low annealing temperature. Only the oligonucleotide 'tailed' DNA generated in the initial cycles is amplified in the later high annealing-temperature cycles. The paints generated using this primer 'paint' the chromosome evenly along its length, although the centromere does not usually label.

Protocol 4

DOP-PCR amplification of sorted chromosomes

The method described below is a modified protocol of Telinius *et al.* (11).

Equipment and reagents

- 10 × DOP buffer: 20 mM MgCl₂, 500 mM KCl, 100 mM Tris–HCl, 1 mg/ml gelatin, pH 8.4
- dNTPs working solution (Pharmacia Ultrapure, supplied as 100 mM solutions) of each dNTP
- *Taq* polymerase (AmpliTaq, Perkin-Elmer-Cetus)
- 6-MW DOP-PCR primer working solution at 1 µg/µl (5′ CCG ACT CGA GNN NNN NAT GTG G 3′; where N = A, C, G, or T in approximately equal proportions).
- Mineral oil (Sigma)
- Thermal cycler
- 1.5% agarose gel and electrophoresis equipment and reagents

A. Primary amplification

1. Combine the reagents in one tube to produce a final concentration of: 1 × DOP buffer, 200 µM each dNTP, 15 µg/ml 6-MW, and 25 units/ml AmpliTaq. Aliquot into PCR tubes either 50 µl or 100 µl per tube.

2. Sort chromosomes directly into the tubes, 500–1000 chromosomes for a 100 µl reaction seems optimal.

3. Overlay with 50 µl of mineral oil.

4. Place the tubes in the PCR machine and execute the following program:
   (a) initial denaturation 5 min at 94°C, followed by:
   (b) 7 cycles of 30 sec at 95°C, 1.5 min at 30°C, 3 min at 72°C with the transition from 30 to 72°C taking 3 min, followed by:
   (c) 35 cycles of 30 sec at 95°C 1 min at 56°C and 3 min 72°C with an addition of 1 sec per cycle to the extension step;
   (d) follow the last cycle with a further 10 min at 72°C.

B. Secondary amplification

1. Combine the reagents in one tube to produce a final concentration of: 1 × DOP buffer, 200 µM each dNTP, 20 µg/ml 6-MW, and 25 units/ml of AmpliTaq—exactly as in the primary amplification except add 10 µl/ml of a previously amplified sample. Include labelled dNTPs if desired. Dispense 50 µl or 100 µl into PCR tubes and overlay with 50 µl of mineral oil.

2. Place the tubes in the thermal cycler and execute a program with the following characteristics:

   (a) initial denaturation 5 min at 95 °C, followed by:

   (b) 10–25 cycles (10–15 cycles is enough if the yield of the primary amplification is good) of 1 min at 95 °C, 1.5 min at 56 °C, and 4 min at 72 °C, with an additional 1 sec per cycle to the extension step;

   (c) follow the last cycle with a further 10 min at 72 °C.

3. Visualize the PCR products by running on a 1.5% agarose gel (the products should be in the size range of 300 bp to 3 kbp). For a suitable DNA standard use the PhiX174 *Hae*III markers.

5.2 Alu-polymerase chain reaction

It may sometimes be desirable to generate paints using primers directed against repeat sequences such as Alu repeats. Alu-primers were originally used to generate paints from human chromosomes in somatic cell hybrids where only the human material is amplified. Alu repeats occur on average about every 4 kb in the genome, but they are not evenly spaced. Painting using Alu paints results in uneven painting along the length of the chromosome giving a pattern corresponding to R-banding. There is a risk that the region of interest will not be amplified when Alu paints are used; on the other hand, the DOP-PCR paints do not paint entirely evenly as the centromere is rarely painted. Alu paints give low background and the pattern along the length of the chromosome can aid chromosome identification.

Protocol 5

Alu-PCR of sorted chromosomes

Equipment and reagents

- 10 × PCR buffer: 15 mM MgCl$_2$, 500 mM KCl, 100 mM Tris–HCl, 1.0% Triton X-100, 1 mg/ml gelatin, pH 8.9
- dNTPs (see *Protocol 4*)
- BK33 primer (5′ CTGGGATTACAGGC-GTGAGC 3′), working solution at 1 µg/µl

- *Taq* polymerase (AmpliTaq, Perkin–Elmer–Cetus)
- Mineral oil (Sigma)

Method

1. Combine the reagents in one tube to give a final concentration of: 200 µM each dNTP, 1 × PCR buffer, 10 µg/ml BK33, and 25 units/ml *Taq* polymerase. Aliquot either 100 µl or 50 µl into PCR tubes.

Protocol 5 continued

2. Sort 500–1000 chromosomes directly into each PCR tube leaving one tube free for the control. (Ideally, the control tube should contain the appropriate amount of sheath buffer.)

3. Execute the following PCR program:

   (a) 2 min initial denaturation at 95 °C, then:

   (b) 35 cycles of 30 sec at 95 °C, 1 min at 55 °C, and 4 min at 68 °C;

   (c) finally extend for 10 min at 68 °C.

4. Label the PCR products either by re-amplifying 5 μl of the PCR products with a labelled dNTP, using 10 more cycles of the same protocol, or by using a nick translation kit.

References

1. Gray, J. W., Carrano, A. V., Moore II, D. H., Steinmetz, L. L., Minkler, J., Mayall, B. H., Mendelsohn, M. L., and Van Dilla, M. A. (1975). *Clin. Chem.*, **21**, 1258.

2. Gray, J. W., Carrano, A. V., Steinmetz, L. L., Van Dilla, M. A., Moore II, D. H., Mayall, B. H., and Mendelsohn, M. L. (1975). *Proc. Natl Acad. Sci. USA*, **72**, 1231.

3. Young, B. D., Ferguson-Smith, M. A., Sillar, R., and Boyd, E. (1981). *Proc. Natl Acad. Sci. USA*, **78**, 7727.

4. Krumlauf, R., Jeanpierre, M., and Young, B. D. (1982). *Proc. Natl Acad. Sci. USA*, **79**, 2971.

5. Trask, B., van den Engh, G., and Gray, J. W. (1989). *Am. J. Hum. Genet.*, **45**, 753.

6. Mefford, H., van den Engh, G., Friedman, C., and Trask, B. J. (1997). *Hum Genet.*, **100**, 138.

7. Stepanov, S. I., Konyshev, V. N., Kotlovanova, L. V., and Roganov, A. P. (1996). *Cytometry*, **23**, 279.

8. Nizetic, D., Zehetner. G., Monaco. A. P., Young, B. D., and Lehrach, H. (1991). *Proc. Natl Acad. Sci. USA*, **88**, 3233.

9. Fawcett, J. J., Longmire, J. L., Martin, J. C., Deaven, L. L., and Cram, L. S. (1994). In *Methods in cell biology*, Vol. 42 (ed. Z. Darzynkiewicz, J. P. Robinson and H. A. Crissman), p. 319. Academic Press, San Diego.

10. Cotter, F. E., Das, S., Douek, E., Carter, N. P., and Young, B. D. (1991). *Genomics*, **9**, 473.

11. Telinius, H., Pelmear, A., Tunnacliffe, A., Carter, N., Behmet, A., Ferguson-Smith, M., Nordenskjöld, M., Pfragner, R., and Ponder, B. (1992). *Genes, Chromosomes and Cancer*, **4**, 257.

12. Suijkerbuijk, R., Matthopoulos, D., Kearney, L., Monard, S., Dhut, S., Cotter, F., Herbergs, J., Gerts van Kessel, A., and Young, B. (1992). *Genomics*, **13**, 355.

13. Stubblefield, E., Cram, S., and Deaven, L. (1975). *Exp. Cell. Res.*, **94**, 464.

14. Sillar, R. and Young, B. D. (1981). *J. Histochem. Cytochem.*, **29**, 74.

15. Matsson, P. and Rydberg, B. (1981). *Cytometry*, **1**, 369.

16. Yu, L. C., Aten, J., Gray, J., and Carrano, A. V. (1981). *Nature*, **293**, 154.

17. Minoshima, S., Kawasaki, K., Fukuyama, R., Maekawa, M., Kudoh, J., and Shimizu, N. (1990). *Cytometry*, **11**, 539.

18. Van den Engh, G. J., Trask, B., Gray, J. W., Langlois, R. G., and Yu, L. C. (1985). *Cytometry*, **6**, 92.

19. Weber, B. H. F., Stöhr, H., Siedlaczck, I., Longmire, J. L., Deaven, L. L., Duncan, A. M. V., and Riess, O. (1994). *Chromosome Res.*, **2**, 201.

Chapter 13

Intracellular ionized calcium, magnesium, membrane potential, and pH

Peter S. Rabinovitch* and Carl H. June[†]

*University of Washington, Department of Pathology, Box 357705, Seattle WA 98195, USA

[†]Department of Molecular and Cellular Engineering, University of Pennsylvania School of Medicine, PA 19104, USA

1 Intracellular ionized calcium

1.1 Introduction

Calcium has an important role as a mediator of transmembrane signal transduction, and elevations in intracellular ionized calcium concentration ($[Ca^{2+}]_i$) are part of the regulation of diverse cellular processes. Measurement of $[Ca^{2+}]_i$ in living cells is thus of considerable interest to investigators over a broad range of cell biology.

Eukaryotic cells in their resting state maintain an internal calcium ion concentration that is far below that of the extracellular environment, through the regulation of calcium pumps and channels within the plasma membrane and intracellular storage of insoluble calcium. Calcium influx is thought to be initiated by membrane depolarization, which opens voltage-gated channels, or by the binding of ligands to receptor-operated channels. In the latter case, the binding of agonist to its specific membrane receptor causes the activation of phospholipase C which causes the hydrolysis of a membrane phospholipid, phosphatidylinositol 4,5-bisphosphate (PIP2) to yield inositol 1,4,5-trisphosphate (IP3) and 2-diacylglycerol (DAG). IP3 or IP3 metabolites then cause the release of calcium from intracellular stores, while DAG and elevated $[Ca^{2+}]_i$ activate protein kinase C. Ca^{2+} and activated protein kinases then continue the cascade of diverse cellular effects. Recent studies have suggested that global intracellular Ca^{2+} signals arise from the summation and coordination of subcellular elementary release events termed 'Ca^{2+} puffs' (1).

Until the early 1980s, measurement of $[Ca^{2+}]_i$ was restricted to large invertebrate cells where the use of microelectrodes was possible. Subsequently, bioluminescent indicators such as aquorin, a calcium-activated photoprotein, were described, but were limited in their application by the necessity for loading

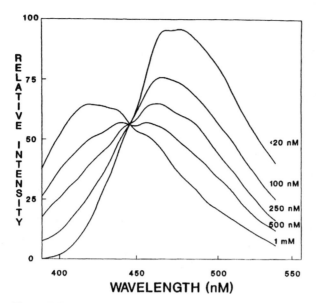

Figure 1 Emission spectra of indo-1 as a function of calcium concentration. At wavelengths below 400 nm, emission in the absence of Ca^{2+} drops to a few per cent of that seen in the presence of Ca^{2+}. Above 500 nm, there is over a fourfold converse difference. The [Ca^{2+}] of EGTA buffer solutions used is shown at the right.

them into cells by microinjection or other membrane disruption. [Ca^{2+}]$_i$ was first measured conveniently in diverse cell types by the development of quin2 (2). The acetoxymethyl ester of the dye is uncharged and diffuses freely into the cytoplasm. Within the cytoplasm it is a substrate for esterases, these enzymes hydrolyse the dye to the tetra-anionic form which is trapped inside the cell (3). Unfortunately, quin2 has a relatively low extinction coefficient and quantum yield, which made detection of the dye at low concentrations by flow cytometry difficult, and at higher concentrations, quin2 itself buffers the [Ca^{2+}]$_i$. In 1985, the first of a new generation of [Ca^{2+}]$_i$ indicator dyes was developed which overcame most of these difficulties (4). One of these dyes, indo-1 ((1-(2 amino-5-(carboxylindol-2-yl)-phenoxy)-2–2'-amino-5'-methylphenoxy) ethane *N,N,N'N'*-tetraacetic acid), was especially well suited for flow cytometry, and showed large changes in fluorescent emission wavelength upon calcium binding (see *Figure 1*).

As described below, use of the ratio of intensities of fluorescence at two wavelengths allows the calculation of [Ca^{2+}]$_i$ independent of variability in intracellular dye concentration. Since indo-1 requires UV excitation, considerable efforts have been devoted to developing visible light-excited probes. Although a single, visible light-excited ratiometric dye has yet to be discovered, several alternatives are available that have proven useful (see *Table 1*). Non-ratiometric dyes that exhibit increased fluorescence upon binding of calcium include fluo-3, rhod-2, and a series of dyes developed by Molecular Probes that include fluo-4, Oregon Green 488 BAPTA, Calcium Green, Calcium Orange, Calcium Crimson,

Table 1 Calcium indicator dyes useful for flow cytometric applications

Indicator	Emission response to elevated calcium	Excitation wavelength (nm)	Emission(s) wavelength (nm)	Calcium affinity K_d (nm)	
				22°C	37°C
Indo-1	Ratio change	325–360	390/520	—	~250
Fluo-3	Increase	488	526	~325*	~800
Fluo-4	Increase	488	516	~345*	—
Oregon Green 488 BAPTA-1	Increase	488	523	~170*	—
Calcium Green-1	Increase	488	530	~190*	—
Calcium Orange	Increase	550	575	~185*	—
Calcium Crimson	Increase	590	610	~185*	—
Fura Red	Decrease	488	660	~400	—
Fluo-3/Fura Red	Ratio change	488	530/660	~400	—

*Molecular Probes, http://www.probes.com, August 1999.

and additional rhodamine-based indicators (5, 6). Calcium Orange, Calcium Crimson, and the rhodamine-based dyes are excited between 550 and 590 nm and are generally used more by image than flow cytometry. Fluo-4 is a recently developed derivative of fluo-3, in which the two chlorine substituents are replaced by fluorine; this results in improved excitation at 488 nm (5). A similar strategy was used to shift the excitation spectra of the fluorine-substituted Oregon Green 488 BAPTA from that of its chlorinated analogue Calcium Green. Fura Red exhibits *decreased* fluorescence intensity upon calcium binding, and as described below, can be combined with the simultaneous cellular loading of fluo-3 to provide a form of ratiometric analysis.

Protocol 1

Loading of cells for calcium analysis with indo-1, fluo-3, fluo-4, or fluo-3 plus Fura Red

Reagents

- Indo-1, fluo-3 (or fluo-4), or fluo-3 and Fura Red acetoxy-methyl esters (Molecular Probes)
- FITC- or PE-conjugated antibody (optional)
- DMSO and/or bleach

Method

1. Incubate cells (<2 × 10⁷/ml) in medium with 1–3 μM indo-1, fluo-3, or fluo-4 or with 4 μM fluo-3 and 10 μM Fura Red (all as acetoxy-methyl esters) at 37°C for 20–40 min. Optimize the probe concentration and loading duration for each cell type.

Protocol 1 continued

2. Incubate aliquots of the above with saturating concentrations of FITC- or PE-conjugated antibody or FITC-conjugated antibody (with indo-1) at 22 °C for 30 min (optional step). Note that antibodies should be azide-free.

3. Centrifuge (100 g, 10 min, 20 °C) and resuspend in medium at the desired cell concentration (usually ~10^6/ml) at 22 °C.

4. Warm aliquots to 37 °C at least 5 min before analysis. Analyse at 37 °C. Ensure that the medium used contains approximately 1 mM Ca^{2+} and that the sheath fluid is saline or buffered saline.

5. Calibrate using one of the techniques described in Section 1.4. When using calcium ionophores, be scrupulous (this is a necessity) in removing residual material before proceeding to the next sample by washing the input tubing and exposed surfaces with DMSO and/or bleach. Note that Br-A23187 may be more easily removed than ionomycin.

1.2 Loading of cells

The uptake and retention of indicator dyes is facilitated by the use of the acetoxymethyl esters of each dye, using the scheme first described for quin2. Cells can be incubated with the permeant form of the dye at 37 °C for 40 min (see *Protocol 1*), under which conditions peripheral blood lymphocytes (PBL) will subsequently typically show approximately 20% of the total dye trapped within the cell (7). After loading, the extracellular dye should be diluted 10- to 100-fold before flow cytometric analysis (7). One incidental benefit of this family of probes is that, like the more familiar use of fluorescein diacetate or carboxy-fluorescein diacetate (see Chapter 15, Section 2), the procedure discriminates between live and dead cells. The latter will not retain the permeant dye and will be excluded by gating during subsequent flow cytometric analysis.

The lower limit of useful intracellular loading concentrations of indicator dyes is determined by the fluorescence detection sensitivity of the flow cyto-meter, and the upper limit is determined by avoiding the buffering of $[Ca^{2+}]_i$ by the dye itself. Fortunately, indo-1, fluo-3 or fluo-4, Calcium Orange, and Calcium Crimson exhibit excellent fluorescence characteristics (30-fold greater quantum yield than quin2 at a given dye concentration (4)) and useful ranges of dye loading are in the few micromolar range. For human peripheral blood T cells, we have found adequate detection with indo-1 or fluo-3 at or above 1 μM load-ing concentrations for cells incubated at 37 °C for 40 min, conditions that achieve ~5 μM intracellular dye concentrations. Buffering of $(Ca^{2+})_i$ was observed as a slight delay in the rise in $[Ca^{2+}]_i$ and a retarded rate of return of $[Ca^{2+}]_i$ to baseline values seen at loading concentrations above 5 μM, and a reduction in peak $[Ca^{2+}]_i$ seen at even higher indo-1 concentrations (7). Murine B cells are more sensitive to calcium indicator-dye buffering and Chused *et al.* (8) recommended a loading concentration of no greater than 1 μM of indo-1. Rates of loading of the dye ester vary between cell types, especially as a consequence of variations in intracellular esterase activity, and more rapid loading rates are seen in platelets and monocytes than in lymphocytes. Rates of loading also vary between different calcium indicator dyes, inversely related to their molecular weight (9); for this reason Oregon Green 488 BAPTA-2 (which contains two dye molecules per BAPTA chelator) is less readily loaded than Oregon Green 488 BAPTA and most other dyes.

The choice of medium in which the cell sample is suspended for analysis can be dictated primarily by the metabolic requirements of the cells, subject only to the presence of millimolar concentrations of calcium (to enable calcium agonist-stimulated calcium influx) and reasonable buffering. The use of phenol red as a pH indicator does not impair the detection of indo-1 or fluo fluorescence signals. Although these Ca^{2+} indicator dyes are not highly sensitive to small fluctuations of pH over the physiological range (4), unbuffered or bicarbonate-buffered solutions can impart uncontrolled pH shifts. If analysis of the release of Ca^{2+} from intracellular stores, independent of extracellular Ca^{2+} influx is desired, the addition of 5 mM EGTA to the cell suspension (final concentration) will reduce Ca^{2+} from several mM to ~20 nM, thus abolishing the usual extracellular to intracellular gradient.

The dye-loaded cells are introduced into the flow cytometer in the usual fashion. Usually, a baseline analysis of unstimulated cells is made, followed by addition of an agonist to the cells and re-analysis. In many experiments, the entire time-course of the experiment will be analysed by flow cytometry, with one of the acquisition parameters being time (discussed more fully below). In such experiments, it is especially important to regulate the temperature of the sample of cells. The generation of a cellular $[Ca^{2+}]_i$ response is an active process and is highly temperature-dependent in most cell types. Thus, the sample chamber must usually be kept warm, and the time that cells spend in cooler tubing in transit to the flow cell must be kept to the minimum (less than 10 seconds), or the sample tubing must also be kept warm. The agonist is ordinarily introduced into the sample by quickly stopping the flow, removing the sample container, adding the agonist, and restarting the flow, 'boosting' the new sample to the flow cell quickly. With practice, this procedure can be completed in less than 20 seconds. If more rapid $[Ca^{2+}]_i$ transients after agonist addition must be analysed, then various agonist injection methods may be utilized so that disruption of the sample flow is not required. A commercial device is available for on-line addition of the agonist using a syringe (Cytek Development).

1.3 Optical considerations

The most commonly used visible excited dyes may be excited by 488 nm argon-laser emission, as shown in *Table 1*. For indo-1, the absorption maximum is between 330 and 350 nm, depending upon the presence of calcium (4); this is well-suited to excitation at either 351–356 nm from an argon-ion laser, or to 337–356 nm excitation from a krypton-ion laser. With the usual emission filters (see below) and the optical efficiency of commercial instruments, laser power requirements are modest: 10–20 mW is satisfactory. Although 325 nm emission from a helium–cadmium laser is slightly below optimal excitation, very good results can be obtained in practice (10).

For the non-ratiometric dyes listed in *Table 1*, an emission filter with a band-pass near the indicated wavelength may be chosen. For ratiometric dyes, the choice requires slightly more consideration. With indo-1, for example, bandpass filters could be chosen to be centred on the 'violet' peak emission of the calcium-bound indo-1 dye (400 nm) and free indo-1 dye 'blue' emission (485 nm). However, wavelengths nearer the isobestic point do not exhibit as large a dependence upon calcium binding. Thus, as shown in *Table 13.2*, the ratio of 475 nm to 405 nm emission in the presence of Ca^{2+} (R_{max}), the ratio in the absence of Ca^{2+} (R_{min}), and the ratio for resting lymphocyte $(Ca^{2+})_i$ (R) yield a maximal Ca^{2+} related change in ratio (R_{max}/R_{min}) of 23.8, and the maximum fluorescence shift that can be observed from initially resting cells (R_{max}/R) is 5.8. At ratios of 495 nm to 395 nm, the changes in ratio are greater, and at wavelengths 530 nm and 395 nm, R_{max}/R_{min} has increased to 82.7 and R_{max}/R is 9.02. Thus, a larger dynamic range in the ratio of wavelengths is obtained if 'blue' emission below 485 nm is not collected and the 'blue' emission bandpass is moved upward above 485 nm, or even into the green, above 500 nm. Similarly, the 'violet' bandpass filter should be chosen to minimize the collection of wavelengths above 400 nm. The choice of a dichroic mirror to separate the emitted fluorescence into the two detectors is not critical; any cut-off wavelength well separated from the 400 nm and 485 or 500 nm ranges will prove satisfactory. For the combination of fluo-3 and Fura Red (see Section 1.6 and *Figure 1*), a standard 'narrow' 530 nm FITC emission filter may be used for fluo-3 detection (if fluo-4 is used, the emission filter should pass 516 nm light). A 675 nm bandpass is optimal to collect Fura Red emission.

A typical optical arrangement for flow cytometry with indo-1 is shown in *Figure.3*. As described subsequently, provision can also be made for detecting fluorochromes used for the analysis of immunofluorescence.

1.4 Calibration to $[Ca^{2+}]_i$

The determination of $[Ca^{2+}]_i$ by measurement of non-ratiometric dyes is sensitive to cell size and changes in intracellular dye concentration as well as

Table 2 Indo-1 ratio shifts are dependent upon filter choices

'Violet' wavelength (nm)	'Blue-green' wavelength (nm)	R_{max}	R_{min}	R	R_{max}/R_{min}	R_{max}/R
405	475	2.38	0.100	0.41	23.8	5.8
405	495	3.59	0.119	0.50	30.2	7.17
395	495	3.51	0.048	0.43	73.1	8.2
395	530	9.68	0.117	1.07	82.7	9.02

By spectrofluorimetry, 2-nm slit width, uncorrected.

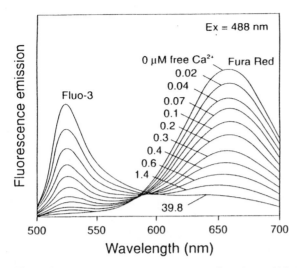

Figure 2 Fluorescence emission spectra of a mixture of fluo-3 and Fura Red in the presence of buffers containing various concentrations of free calcium, as noted. The Fura Red concentration is approximately 10 times that of fluo-3. (Courtesy of Molecular Probes, Inc., with copyright permission.)

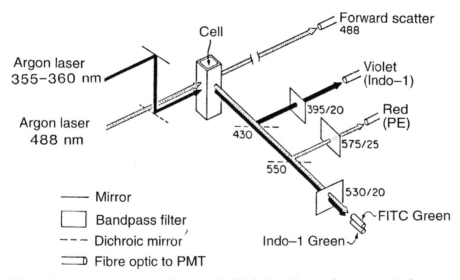

Figure 3 An optical arrangement for analysis of indo-1 and immunofluorescence by flow cytometry. On many instruments, the fibre-optic cables would be replaced by the photomultiplier tubes directly.

$[Ca^{2+}]_i$. In this circumstance, accurate calibration requires measurements at the end of each set of assays, by determination of the fluorescence intensity using ionophores and buffered $[Ca^{2+}]$ standards (see Section 1.4.2). In contrast, with indo-1, the optical properties of the dye can be used to perform a simpler calibration.

209

1.4.1 Calibration of ratiometric dyes based on Ca-binding properties

For ratiometric dyes, the use of the $[Ca^{2+}]$-dependent shift in dye emission wavelength allows the ratio of fluorescence intensities of the dye at the two wavelengths to be used to calculate $[Ca^{2+}]$:

$$[Ca^{2+}] = K_d \times \frac{(R - R_{min})}{(R_{max} - R)} \times \frac{S_{f2}}{S_{b2}}; \qquad (1)$$

where K_d is the effective dissociation constant (250 nM at 37°C, pH 7.08 and ionic strength 0.1), R, R_{min}, and R_{max} are the fluorescence intensity ratios of violet/blue fluorescence at resting, zero, and saturating $[Ca^{2+}]_i$, respectively; and S_{f2}/S_{b2} is the ratio of the blue fluorescence intensity of the calcium-free and bound dye, respectively (4). Because this calibration is independent of total intracellular dye concentration and instrumental variation in efficiency of excitation or emission detection, it is not necessary to measure the fluorescence of the dye in the calcium-free and saturated states for each individual assay. In principle, it is sufficient to calibrate the instrument once after set-up and tuning by measurement of the constants R_{max}, R_{min}, S_{f2}, and S_{b2}, after which only R is measured for each subsequent analysis on that occasion.

The values of R_{max} and R_{min} for indo-1 can be obtained by lysing the cells in order to release the dye to determine fluorescence at zero and saturating $[Ca^{2+}]_i$, as is carried out with quin2. However, this is not possible with flow cytometry, due to the loss of cell-associated fluorescence. Another strategy is the use of an ionophore to saturate or deplete $(Ca^{2+})_i$, to allow approximation of the true endpoints without cell lysis. For this approach the ionophores ionomycin and Br-A23187 are best suited, due to their specificity for calcium and low fluorescence. When flow cytometric quantitation of fluorescence from intact cells treated with ionomycin or ionomycin plus EGTA was compared with spectrofluorimetric analysis of lysed cells in medium with or without EGTA, the indo-1 ratio of unstimulated cells (R) and the ratio at saturating $[Ca^{2+}]$, R_{max}, were similar by both techniques (7). The latter indicates that ionomycin-treated cells reach near-saturating levels of $[Ca^{2+}]_i$. The value of R_{min} that can be obtained by treating cells with ionomycin in the presence of EGTA, however, is substantially higher than either that predicted from the spectral emission curves (see *Figure 13.1*) or that obtained by cell lysis and spectrofluorimetry. Spectrofluorimetric quantitation with either quin2 or indo-1 indicates that $[Ca^{2+}]_i$ remains at approximately 50 nM in intact cells treated with ionomycin and EGTA.

Due to the inability to obtain a valid flow cytometric determination with calcium-free dye, we have used for calibration the values of R_{min} and S_{f2}/S_{b2} derived from spectrofluorimetry, either of the indo-1 pentapotassium salt or after lysis of indo-1-loaded cells in the presence of EGTA. It is essential that the same optical filters be used for flow cytometry and spectrofluorimetry, since the standardization is very sensitive to the wavelengths chosen. As an alternative, there are several potential schemes for using the flow cytometer to directly analyse the fluorescence of dye in solution; the fluorescence measurement must be converted to a brief pulse for processing by the flow cytometer, either by strobing the exciting laser beam, or by chopping the photomultiplier (PMT) output signals (11). Even if careful calibration of the fluorescence ratio to $[Ca^{2+}]_i$ is not being performed for a particular experiment, ordinary quality control can include a determination of the value of R_{max}/R. Unperturbed cells will usually be found to have a reproducible value of $[Ca^{2+}]_i$, and day-to-day optical variations in the flow cytometer are usually minimal (with the same filter set), thus a limited range of R_{max}/R values should be obtained.

1.4.2 Calibration using calcium buffers

An alternative calibration strategy, suitable for use with all dyes, was first suggested by Chused *et al.* (8), and is based upon the equilibration of intracellular calcium concentrations with external calcium in a series of buffers with carefully prepared, known calcium concentrations. These authors used a cocktail of ionomycin, nigericin, high concentrations of potassium, 2-deoxyglucose, azide, and carbonyl cyanide *m*-chloro-phenylhydrazone to collapse the calcium gradient to zero, thereafter assuming that $[Ca^{2+}]_i = [Ca^{2+}]_0$. Thus, this technique allows one to estimate $(Ca^{2+})_i$ without the need to determine R_{min}, S_{f2}, or S_{b2}, although it is subject to the limitations imposed by the precision with which one can prepare the series of calcium buffers that yield accurate and reproducible free calcium concentrations.

The accurate prediction of $[Ca^{2+}]$ in buffer solutions is dependent upon a variety of interacting factors so that care must be exercised in formulating Ca^{2+} standards. It is impossible to use methods to prepare directly a solution that contains calcium in a concentration similar to that found inside living cells due to the contamination of laboratory water by calcium (generally micromolar amounts). In addition, it is impossible to prepare solutions with sufficient precision using gravimetric methods because $CaCl_2$ and EGTA each contain variable amounts of water. Therefore, a strategy is used that depends upon buffers with known free calcium concentration. The determination of the ionized calcium concentration in an EGTA buffer system is dependent upon the magnesium concentration; other metals such as aluminium, iron, and lanthanum also bind avidly to EGTA. In addition, the dissociation constant of Ca^{2+}–EGTA is a function of pH, temperature, and ionic strength. For example, in a Ca^{2+}/Ca^{2+}–EGTA buffer system, changing the pH from 7.4 to 7.1 can result in the ionized calcium changing from 110 nM to 375 nM. It is also important to prepare the buffers using the 'pH metric technique' (12) because of the varying purity of commercially available EGTA. Because of these complexities, most users will wish to use commercially available buffer solutions for calibration (Molecular Probes).

Protocol 2, described below, employs a series of calcium buffer solutions, purchased or precisely prepared as above, to perform an *in-situ* calibration in which the cellular cytosol is maintained at a particular calcium concentration for the standardization of indicator dye fluorescence (or fluorescence ratio). To create solutions of known calcium concentration, binary mixtures are prepared, consisting of buffer A (an equimolar solution of calcium and a calcium chelator such as EGTA) and buffer B, (identical to buffer A except that it lacks calcium). EGTA or BAPTA are used because they have high selectivity over magnesium, an ion that is $\sim 10^4$ times more abundant than calcium in the cytoplasm, and therefore can be used to control calcium in the presence of physiological concentrations of magnesium. To obtain a buffer with a desired calcium concentration, buffers A and B are mixed at the necessary ratio for a given temperature, pH, magnesium concentration, and ionic strength.

Protocol 2

Calibration of $[Ca^{2+}]_i$ using calcium buffers

Reagents

- Calcium calibration buffer concentrates: 100 mM K_2H_2-EGTA and 100 mM K_2Ca-EGTA (Molecular Probes, cat no. C-3723)

- 'Poisoned' Dulbecco's phosphate-buffered saline (DPBS; approximately 33 μM Ca^{2+}): DPBS with 20 mM Hepes pH 7.20 without calcium or magnesium containing the following cellular poisons: ionomycin (1 mg/ml stock solution in DMSO to 3 μg/ml final); nigericin (10 mg/ml stock solution

in methanol to 2.0 μg/ml final); carbonyl cyanide m-chlorophenylhydrazone (CCCP, 1 mM stock solution in DMSO at 10 μM final); 2-deoxyglucose (1 M stock solution in water, 40 mM final); sodium azide (3 M stock solution in water, 60 mM final)

- Indo-1, fluo-3 (or fluo-4), or fluo-3 and Fura Red acetoxy-methyl esters (Molecular Probes)

A. Preparation of calcium:EGTA buffers[a]

1. Buffer A (zero Ca^{2+}). Add 10.00 ml of 100 mM K_2H_2-EGTA to 90.00 ml 'poisoned' DPBS.

2. Buffer B (~35 μM Ca^{2+}). Add 10.00 ml of 100 mM K_2Ca-EGTA to 90.00 ml 'poisoned' DPBS.

3. Buffer 2 (31 nM Ca^{2+} in an intracellular environment anticipated to contain ~1 mM Mg^{2+}). Mix 40.00 ml of buffer A plus 10.00 ml buffer B.

4. Buffer 3 (74 nM Ca^{2+}). Mix 36.00 ml of buffer 1 plus 10.00 ml buffer B.

5. Buffer 4 (135 nM Ca^{2+}). Mix 32.00 ml of buffer 2 plus 10.00 ml buffer B.

6. Buffer 5 (246 nM Ca^{2+}). Mix 28.00 ml of buffer 3 plus 12.00 ml buffer B.

7. Buffer 6 (462 nM Ca^{2+}). Mix 24.00 ml of buffer 4 plus 14.00 ml buffer B.

8. Buffer 7 (876 nM Ca^{2+}). Mix 20.00 ml of buffer 5 plus 18.00 ml buffer B.

B. Calibration

1. Load cells with indo-1, fluo-3 or fluo-4, or fluo-3 plus Fura Red in buffer A.

2. Wash the cells once in buffer A with dye (there must be no residual serum or other components that will change the calcium concentration of the buffer) and resuspend aliquots of 5×10^5 cells in 1 ml of each of the buffers above.

3. Incubate cells at 37°C for 90 min to permit equilibration of intracellular calcium concentration in the 'poisoned' cells to the extracellular calcium concentration.

4. Analyse the cells to determine the steady-state fluorescence ratio (for indo-1 or fluo-3 plus Fura Red-loaded cells) or absolute fluorescence (for fluo-3 or fluo-4 loaded cells) of cells in each buffer solution. Exclude dead cells with poor dye retention.

5. Plot the peak ratio channel (for indo-1 or fluo-3 plus Fura Red-loaded cells) or channel (for fluo-3 or fluo-4 loaded cells) vs. calcium of each buffer. This calibration curve may be used for subsequent experiments using the same cells and loading conditions.

[a] For all buffers use plasticware rather than glass.

1.5 Data analysis and display

Software is available, either from the cytometer manufacturer or third-party suppliers, to facilitate the analysis and presentation of kinetic data obtained from the measurement of intracellular ions. The most elementary form of the display of indo-1 fluorescence on the flow cytometer is as a bivariate plot of 'violet' vs. 'blue' signals. In this case, the increase in ratio seen with increased $[Ca^{2+}]_i$ will be observed as a rotation around the axis through the origin. This method of ratio analysis is cumbersome, however, and fortunately commercial flow cytometers all have some provision for a direct calculation of the value of the fluorescence ratio of 'violet'/'blue' itself, either by analog circuitry or by digital computation. By either type of calculation, it is important that no artefactual offset be introduced in the ratio; this can be quickly tested by altering the excitation power over a broad range of values in an analysis of a non-perturbed indo-1-loaded cell population—a correctly calculated ratio will show no dependence upon excitation intensity. Alternatively, it can be shown that loading cells with a broad range of indo-1 concentrations should result in a constant value of the 'violet'/'blue' ratio (7).

The most informative and elegant display is obtained by a bivariate plot of ratio vs. time. For non-ratiometric dyes, such as fluo-3 or fluo-4, the analogous plot is that of fluorescence intensity vs. time. The bivariate data can be viewed in real time or subsequently plotted as a 'dot plot' (see Figure 4A). Alternatively, the data can be displayed as an 'isometric plot' in which the x-axis represents time, the y-axis $[Ca^{2+}]_i$, and the z-axis, the number of cells (see Figure 4B). Kinetic changes in $[Ca^{2+}]_i$ are easily visible in these data presentations, and the time resolution of such changes is limited only by the number of channels available on the 'time' axis. The mean magnitude of the response, heterogeneity among the responding population, and changes in the fraction of cells responding to a particular stimulus are easily observed with these displays. For example, it can be seen in Figure 4 that not all mouse thymocytes respond to stimulation with anti-CD3, and of those which do, the $[Ca^{2+}]_i$ elevations are quite variable. By the use of simultaneous immunofluorescence with indo-1 analysis (discussed

213

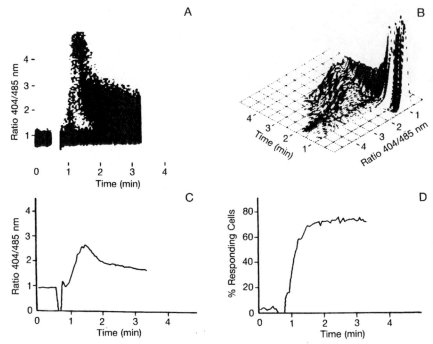

Figure 4 Display of indo-1 temporal analysis results. Indo-1-loaded murine thymocytes (C57BL/6) were stimulated with anti-CD3 antibody 145–2C11 40 seconds after the start of analysis. In panel A the results are displayed in a 'dot plot' in which the value of the indo-1 violet/blue ratio is plotted for each cell in a 100 by 100 bivariate array vs. time. The number of cells in a particular pixel is shown by the intensity of the display at that point, which in the original display ranged over 16 grey levels (with considerable loss in the reproduction and printing process). In panel B the same data is displayed as an isometric plot in which the z-axis represents the number of cells with a particular indo-1 ratio at that time interval. In panels C and D the data is reduced to univariate plots of the mean indo-1 ratio vs. time, and the per cent of 'responding' cells (i.e. the proportion with ratios which are two standard deviations above the mean of the resting cell distribution), respectively (analysed with MultiTime from Phoenix Flow Systems).

below), some of this heterogeneity can be shown to be due to differences between cell subsets (7).

For a quantitative comparison between different cell treatments, and a quicker grasp of data trends, it is often useful to distil the bivariate data into a single descriptive parameter vs. time. Calculation of the mean y-axis value for each x-axis time interval (see *Figure 4C*) allows the data to be presented as a mean ratio or mean intensity vs. time, or after calibration, as the mean $[Ca^{2+}]_i$ of the population vs. time (7). This mode of data presentation is well established in the calcium literature from bulk measurements by spectrofluorimetry. While this presentation yields much of the information of interest, data relating to the heterogeneity of the $(Ca^{2+})_i$ response is lost. Some of this missing information can be displayed by a calculation of the proportion of cells which 'respond' to a stimulus: a threshold value of the resting distribution of fluorescence ratios is

214

chosen at which only 5% (for example) of the control cells are above this value. The proportion of cells responding by ratio elevations above this threshold vs. time yields a presentation that is informative of the heterogeneity of the response (see *Figure 4D*). Not all the information contained within the original bivariate data is contained within either or both of these univariate displays, however. For example, in *Figure 4*, panels A and B, it is apparent that the earliest cell response consists of a small subset of the cells (approximately 10%) which attain a very high magnitude response for a brief interval of time. The subsequent response by approximately 50% of cells is of a lower magnitude; however, because a larger proportion of cells participate in this phase, the later, lower response is that which is associated with the peak mean $[Ca^{2+}]_i$ response (panel C) and maximum per cent responding cells (panel D). The early response of fewer cells to high $[Ca^{2+}]_i$ values would thus not be apparent in the univariate data. Studies of several cell types have shown that the increasing mean response of a population of cells is accounted for by a larger proportion of responding cells, but that the maximal calcium response of individual responding cells at any given dose of stimulatory ligand is not related to the degree of receptor occupancy (13, 14).

1.6 Fluo-3/Fura Red for ratiometric analysis using 488 nm excitation

Indo-1 has been the ratiometric calcium indicator dye most commonly used in flow cytometry because of its shift in emission frequency when excited at a single wavelength. A significant drawback to the use of indo-1, however, is the requirement for UV excitation, which is not as widely available as 488 nm excitation. A ratiometric analysis with visible excitation is possible using two dyes simultaneously: Fura Red (15) and fluo-3, a fluorescein-based, calcium-sensitive probe (16). As seen in *Table 1*, both dyes are excited at 488 nm; but fluo-3 fluoresces in the green region with increasing intensity when bound to calcium, while Fura Red exhibits the inverse behaviour, fluorescing most intensely in the red region when not calcium-bound. Fluo-4 would be expected to behave similarly to fluo-3 when used in this assay. The spectra characteristics of the mixture of fluo-3 and Fura Red (see *Figure 2*) is reminiscent in shape to the UV excited emission of indo-1 (see *Figure 1*). Fura Red emissions are dimmer than emissions from cells loaded with indo-1 and fluo-3 at the same concentration, and approximately 2–3.5 times as much Fura Red as fluo-3 may be required to produce emissions of optimal intensity; loading concentrations of 10 and 4 μM, respectively have been suggested (17). As *Figure.5* shows, the fluo-3/Fura Red ratio provides a better resolved picture of the increase in $[Ca^{2+}]_i$ associated with the cellular response to the antibody stimulus than does either fluo-3 or Fura Red alone. Qualitatively, the fluo-3/Fura Red ratio (see *Figure 5C*) achieves results similar to indo-1 (see *Figure 5D*). There is a larger magnitude response, with less variability in measurements from different cells (see *Table 3*). The narrower distribution of resting ratios from non-stimulated cells can be observed graphically

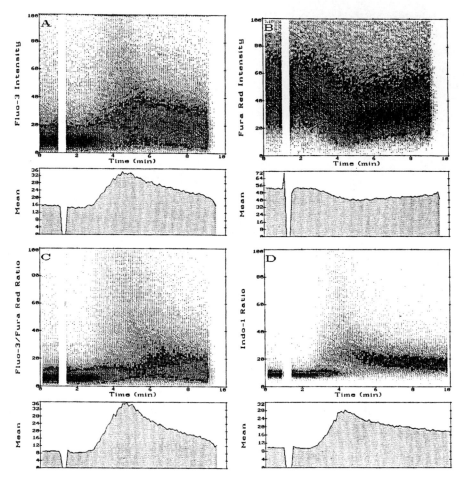

Figure 5 Kinetic displays of fluorescence intensity/ratio and mean values of fluo-3 intensity (A), Fura Red intensity (B), and fluo-3/Fura Red ratio (C) of cells simultaneously loaded with 4 μM fluo-3 + 10 μM Fura Red esters, gated for CD4+ PE-labelled cells. Cells were stimulated with 10 μg/ml anti-CD3 antibody at time 1 min. Cells loaded with 2 μM of indo-1 ester and similarly analysed are shown for comparison (D). (Analysed with MultiTime, Phoenix Flow Systems.)

by comparing the pre-stimulated regions of *Figures 5A* and *5B* with the regions in *Figures 5C* and *5D*. As described in Section 1.7, the fluo-3/Fura Red ratio can also be used simultaneously with analysis of PE-labelled immunofluorescence markers.

1.7 Simultaneous analysis of $[Ca^{2+}]_i$ and other fluorescence parameters

All the calcium probes described above may be combined with immunofluorescence in order to restrict the calcium analysis to a subset of cells of interest. Detailed considerations for each commonly used dye or combination are de-

216

Table 3 Comparison of calcium signals from fluo-3, Fura Red, fluo-3/Fura Red, and indo-1[a]

Indicator	Fold change from baseline ($\sim$300 nM Ca^{2+} response)[b]	SD of resting cells as fraction of response	SD of resting cells as nM Ca^{2+}
Fluo-3 ($n = 4$)	3.5 ± 0.1	0.134	40.8
Fura Red ($n = 4$)	1.5 ± 0.1	0.441	111.3
Fluo/Fura ratio ($n = 4$)	4.6 ± 0.2	0.081	23.5
Indo-1 ratio ($n = 12$)	2.4 ± 0.1	0.076	20.6

[a]Adapted from Novak and Rabinovitch (17).
[b]10 µg/ml CD3 mAb stimulus.

scribed below. In the simultaneous analysis, however, caution should be exercised because the presence of an antibody probe can itself alter the cellular $[Ca^{2+}]_i$. It is increasingly clear that binding of a monoclonal antibody (mAb) to cell-surface proteins can alter $[Ca^{2+}]_i$, even when these proteins are not previously recognized as part of a signal transducing pathway. For example, antibody binding to CD4 will reduce CD3-mediated $(Ca^{2+})_i$ signals in human cells; if instead, the anti-CD4 mAb is cross-linked simultaneously with CD3, as with goat–anti-mouse anti-serum, then the signal is augmented (18). In addition, the epitope of the receptor recognized by the antibody is important, and it is possible to choose antibodies that are either stimulatory or non-stimulatory (see ref. 19 for an example).

As a consequence of this concern, a reciprocal staining strategy should be used whenever possible, so that the cellular subpopulation of interest is un-labelled while undesired cell subsets are identified by mAb staining. The CD4+ subset in PBL may be identified, for example, by staining with a combination of CD8, CD20, and CD11 mAbs (7), and the CD5+ subset can be identified by staining with CD16, CD20 and anti-HLA-DR. Finally, it is important, when staining cells with mAb for functional studies, that the antibodies be azide-free, so that metabolic processes are uninhibited. Commercial antibody preparations may thus require dialysis before use.

Depending upon the calcium-sensitive dye used, there are a number of strategies for the combinations of fluorochromes used:

(a) *Fluo-3 or fluo-4*. Because the emission characteristics of these dyes are virtually identical to those of FITC (16), their use can be combined with almost all of the dyes used with multicolour analysis in addition to FITC.

(b) *Fluo-3/Fura Red ratio*. The emission characteristics of this dye combination (see *Figure 13.2*) have a 'window' between the fluo-3 and Fura Red emission peaks that allows the observation of 488 nm excited PE emission (17). When T lymphocytes, for example, are loaded with 10 µM of Fura Red, emission from Fura Red is approximately 35% as intense as PE–CD4+ emission in the region of 562–588 nm. Using spectral crossover-compensation, this degree of overlap allows the use of bright and moderately bright PE-labelled antibody probes (see *Figure 5*). If a second laser is available, additional label combinations can be used.

(c) *Indo-1*. Although the broad spectrum of indo-1 fluorescence emission precludes the simultaneous use of a second UV-excited dye, the use of a second laser (e.g. 488 nm argon-ion or He–Ne) allows additional information to be derived from dyes exited by visible light.

The flow cytometric assay of intracellular calcium with simultaneous immunofluorescence is only slightly different from the assay of calcium alone. The labelling with fluorescent antibody will ordinarily be performed after labelling with the calcium-sensitive dye (see *Protocol 1*), since the incubation at 37°C with the latter would result in internalization and loss of the antibody label. Room-temperature incubation of the cells (pre-loaded with the calcium indicator dye) with the antibody is recommended, rather than on ice as frequently done for antibody labelling only, since chilled cells can subsequently exhibit altered calcium metabolism, even after re-warming. The antibody will most commonly be labelled with FITC or PE. Excitation and emission wavelengths are those usually employed for these dyes. Note that in many instruments it will be optimal to detect 'blue-green' indo emission using the same 530 nm emission filter (and perhaps the same PMT) as used for FITC. Such a configuration is shown in *Figure 3*. The remainder of the analysis is conducted as usual, except that data from the calcium-sensitive probe is gated upon the immunofluorescence, so that histograms or bivariate plots of indo-1 ratios or fluo-3 intensities (or intensities of other dyes shown in *Table 1*) are presented for 'positive' and 'negative' immunofluorescently labelled cells separately.

As an example of the '3-colour' detection of two indo-1 fluorescence wavelengths and FITC, *Figure 6* shows that after CD2 (alternate pathway) stimulation of peripheral blood lymphocytes, it was found that both CD5+ T cells and a subset of CD5– cells mobilized calcium. Other experiments showed that the CD5– responding cells were CD16+ large granular lymphocytes (LGL), and that after CD3 (antigen-specific) stimulation, more than 90% of T cells responded—as was expected, LGL did not respond (20). As can be seen in *Figure 6*, the pattern of the calcium signal after CD2 stimulation differed in that calcium mobilization in LGL was early in onset and low in magnitude, while in T cells the calcium signal was delayed in onset and high in magnitude.

Combining the analysis of two antibody labels with the calcium-sensitive dye (for example, FITC, PE and indo-1) allows $[Ca^{2+}]_i$ determination in complex immunophenotypic subsets. On instruments with no provision for the analysis of four separate fluorescence wavelengths, *Figure 3* illustrates that detection of both FITC and the higher indo-1 wavelength with the same filter element (but temporally delayed) may allow successful implementation of these experiments. Gating the analysis of the calcium-sensitive dye upon windows of FITC vs. PE fluorescence allows information relating to each identifiable cellular subset to be derived from a single sample. Thus, for example, while '3-colour'

218

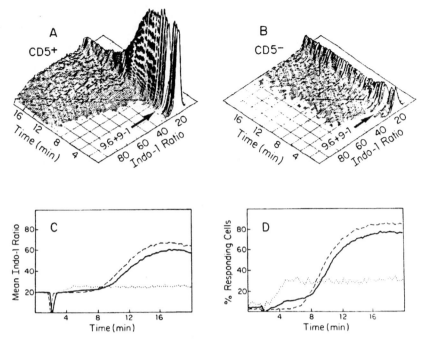

Figure 6 Indo-1 analysis with simultaneous immunofluorescence staining: the effect of CD2 stimulation on the $[Ca^{2+}]_i$ of CD5+ or CD5– lymphocyte subsets. Indo-1-loaded PBL were stained with CD5 (PE-conjugated antibody 10.2), and stimulated with CD2 antibodies 9.6 and 9.1. The indo-1 analyses of PE-positive and -negative cells are displayed in panels A and B, respectively. (C) Time course for the development of the increased mean indo-1 ratio and (D) the percentage of responding cells above a 2 SD threshold ratio (see text) are displayed ($^{×××}$, total cell population analysed; $^{× × ×}$, CD5+ cells; ·····, CD5– cells). (Reproduced from ref. 20, used with copyright permission of the American Society for Clinical Investigation.)

analysis can show that the CD4+ subset shows a more vigorous response to anti-CD3 and lectin stimulation than does the CD8+ subset (7, 21), simultaneous staining of CD4 and CD45R can be used to show that, within the CD4 subset, the $(Ca^{2+})_i$ response to anti-CD3 has a lower magnitude, but is more rapid in onset within the CD45-RO subset (cells with memory function) than in the CD45-RA subset (7).

Finally, we should note that strategies similar to the above may be used with probes for other physiological responses in cells, so that these may be performed simultaneously with $[Ca^{2+}]_i$ measurement. The simultaneous analysis of membrane potential and $[Ca^{2+}]_i$ is discussed later in this chapter. It is also possible to analyse intracellular pH simultaneously with $[Ca^{2+}]_i$ (22, 23).

1.8 Sorting on the basis of $[Ca^{2+}]_i$ responses

Cell sorting based on calcium responses has the following advantages:

(a) Calcium signalling is one of the few biochemical parameters of cell activation that can be measured non-destructively.

(b) Sorting on the basis of calcium dye fluorescence can be an important tool for the selection and identification of genetic variants in the biochemical pathways leading to Ca^{2+} mobilization and cell growth and differentiation.

(c) Cell sorting of indo-1-loaded lymphocytes is a powerful method for isolating antigen-responsive T or B cells from polyclonal populations of cells (24).

The sensitivity of the approach is a function of the magnitude of the calcium signal in the responding cell population, and the uniformity of calcium in the non-responding population. Results of artificial mixing experiments with Jurkat and K562 T-cell and myeloid leukaemia cell lines indicated that subpopulations of cells with variant $[Ca^{2+}]_i$ comprising <1% of total cells could be accurately identified (7). Sorting on the basis of indo-1 or fluo-3 has been used by a number of investigators to identify mutant or variant cells that show a more or less $(Ca^{2+})_i$ response to stimulus (25). This technique is not limited to lymphoid cells, as a specific calcium response was used to sort gastrointestinal cells that secrete cholecystokinin (26).

One of the limitations of this approach is the transient nature of some calcium responses. This has been approached by the development of devices for continuous in-line agonist addition, permitting the sort of a particular time window of a kinetic response (27). Commercial devices of this kind are available (Cytek Development). Another limitation is that calcium responses may consist of sustained responding cells and cells with oscillatory responses; the oscillatory cells may be misclassified as 'non-responders'. Finally, if reproductive viability of the sorted cells is desired, one should minimize UV excitation if indo-1 is used, or use one of the probes that are excited by visible light. In spite of these limitations, sorting on the basis of calcium dye fluorescence can be an important tool for selecting and identifying genetic variants in the biochemical pathways leading to Ca^{2+} mobilization and cell growth and differentiation, and may be used to investigate differences in subpopulations in developmental, differentiated, or ageing variants (28).

1.9 Limitations of $[Ca^{2+}]_i$ assays

The analysis of $[Ca^{2+}]_i$ using indicator dyes assumes that the dye is uniformly distributed within the cytoplasm. In several cell types, fura-2 has been reported to be compartmentalized within organelles. In bovine aortic endothelial cells, fura-2 has been found to be localized to mitochondria, however, indo-1 remains diffusely cytoplasmic. In resting lymphocytes, there is as yet no evidence of the subcellular localization of indo-1, consistent with the concurrence of calibration experiments with predicted results. Compartmentalization may be of greater concern in neutrophils, monocytes, and some established cell lines. In addition, compartmentalization of a probe is enhanced by the prolonged incubation of cells at 37°C, and thus, cells should be stored at room temperature until warmed just before use. If prolonged incubation at 37°C is necessary, then suitable controls should be included to confirm that the behaviour of the dye in response to a standard agonist is unaltered. It is possible that there will be fewer problems with compartmentalization of indo-1 than with fura-2; however, it seems advisable to examine the cellular distribution of indo-1 microscopically, and in each new application to confirm the expected behaviour of the dye using the calibration procedures described previously.

It has been suggested that both fura-2 and indo-1 may be incompletely de-esterified within some cell types. Since the ester has little fluorescence spectral dependence upon Ca^{2+}, the presence of this dye form could lead to false estimates of $[Ca^{2+}]_i$. It has been proposed that since indo-1 fluorescence, but not that of the indo-1 ester, is quenched in the presence of millimolar concentrations of Mn^{2+}, that Mn^{2+} in the presence of ionomycin can be used as a test of the complete hydrolysis of the indo-1 ester within cells. If, for a particular cell type loaded with indo-1, the values of R and R_{max} obtained by flow cytometry are in good agreement with the values obtained by spectrofluorimetry, then it would be unlikely that the dye is in a compartment inaccessible to cytoplasmic Ca^{2+}, in a form unresponsive to $[Ca^{2+}]_i$ (e.g. still esterified) or in a cytoplasmic environment in which the spectral properties of the dye were altered.

Indo-1 has been found to be remarkably non-toxic to cells subsequent to loading. Analysis of the proliferative capacity of either human T lymphocytes (7) or murine B lymphocytes (8) has shown unaltered cell behaviour after loading with indo-1. This is especially pertinent to applications of cell sorting based on $[Ca^{2+}]_i$, as described above.

It is not known to what extent the residual heterogeneity in $(Ca^{2+})_i$ responses represents the effect of oscillations of $[Ca^{2+}]_i$ within individual cells as a function of time; since the same cells cannot be repeatedly examined by flow cytometry, other techniques must be used to address this question. Digital microscopy using fura-2 has shown that, under certain circumstances, B-cell stimulation produces repetitive, transient oscillations of $[Ca^{2+}]_i$ rather than sustained elevations of $[Ca^{2+}]_i$ that are not detectable by flow cytometry (29).

Artefactual alterations in calcium responses may occur for many reasons. Szollosi and co-workers have compared homeostasis in glioblastoma cells by flow cytometry or video microscopy (30). Cells analysed by flow cytometry showed a shorter duration of evoked responses than did attached cells when viewed by microscopy. As noted above, azide can inhibit responses, and ethanol inhibited the mitogen-induced initial increase in $[Ca^{2+}]^i$ in mouse splenocytes (31), so that diluent controls must be performed diligently. Indo-1 has been shown to 'load' into fat droplets, so that artefacts can occur during the analysis of bone marrow cells or adipocytes (32). Finally, both fura-2 and fluo-3 are rapidly inactivated by hydroxyl radicals and enzymatically inactivated by peroxidase/H_2O_2. This results in a decrease in the dynamic range of sensitivity of both dyes to Ca^{2+}, as well as in a decrease in the affinity of fluo-3 for Ca^{2+} (33). Thus, oxidation of calcium probes may alter the interpretation of results, in that the absence of changes in the calcium fluorescence signal can be the result of probe deactivation by free oxygen radicals rather than the lack of actual Ca^{2+} changes.

PETER S. RABINOVITCH AND CARL H. JUNE

2 Magnesium

Magnesium is an intracellular divalent cation that is essential for cellular homeostasis. It is present in millimolar concentrations in cells (34), and approximately half of this is free. A role for $[Mg^{2+}]_i$ in cell signalling is established in the prokaryotic PhoP/PhoQ system. However, while a role has not been firmly established in eukaryotic cells, it has been suggested by multiple investigators—as, for example, by the demonstration of a particular form of phospholipase C that is regulated by magnesium concentration (35). The first useful probe specific for ionized magnesium, magindo-1, was developed by scientists at Molecular Probes. The fluorescence characteristics and use of this probe are similar to indo-1. A number of visibly excited probes have subsequently been developed at Molecular Probes, including Magnesium Green, Magnesium Orange, mag-fluo-4, mag-rhod-2, mag-X-rhod-1, and Mag-Fura Red (36). These dyes can also be used as low-affinity Ca^{2+} indicators in situations where the more sensitive dyes described in Section 1 would be saturated by high $[Ca^{2+}]_i$. Applications for magnesium indicators remain limited but may increase, since, for example, it has been shown that early $(Mg^{2+})_i$ mobilization occurs in some lymphocytes after mitogen stimulation (37).

3 Membrane potential

3.1 Introduction

Resting cells maintain large gradients between intracellular and extracellular concentrations of a variety of ions, including Ca^{2+}, K^+, Na^+, and Cl^-. The relative permeability of the membrane to K^+ ions is greater than that of other ions, and so the leakage of K^+ ions establishes an electron counter-gradient and the cytoplasm becomes electron-negative with respect to the external medium. This K^+ electrochemical gradient is the most significant contribution to the negative membrane potential of most eukaryotic cells. It has been postulated that the maintenance of a large, negative transmembrane potential is a control mechanism to arrest cells in an inactive stage. It has also been suggested that changes in cell membrane potential, which occur in various cell types very quickly after binding of ligands to transmembrane receptors, are mediators of subsequent physiological cellular responses. Detailed investigation of membrane potential has been possible in large cells by direct measurement with implanted microelectrodes. However, our understanding of the role of membrane potential in cell activation of smaller cells has been dependent upon the development and utilization of potential sensitive indicator probes.

3.2 Indicators of membrane potential

3.2.1 General remarks

A charged lipophilic molecule can serve as an indicator of membrane potential, since it is expected to partition between the cell and surrounding medium according to the Nernst equation:

$$\frac{C_i}{C_o} = m^{-nF/RT};$$ (2)

where C_i and C_o are the cytostolic and extracellular indicator concentrations, n is the charge of the indicator (positive for a cation), m is the membrane potential, and F, R, and T are the Faraday, the gas constant, and temperature. For a cationic indicator, the cellular concentration falls as the membrane potential declines towards zero, and rises if the cell hyperpolarizes (i.e. the cytosol becomes more electronegative with respect to the medium). Such a molecule serves as a useful indicator of membrane potential if changes in the partitioning of the indicator are readily detectable (for example, by altered fluorescence intensity of cells which take up or release the molecule according to the membrane potential) and if the indicator does not itself perturb the membrane potential or cause cellular toxicity.

3.2.2 Cyanine dyes

Hoffmann and Laris (38) described a family of cyanine dyes that are useful indicators for cells in solution. These fluorescent dyes have a single negative charge delocalized over an extensive pi-electron system in a highly symmetrical molecule (see Chapter 2, *Figure 3*). The shorthand nomenclature $DiYC_n(2m+1)$ has been introduced for these dyes, where the 'Y' member of the ring structure may be oxygen (O), sulfur (S), or isopropyl (I). The length of the alkyl side chains, 'n,' affects the lipid solubility, and 'm,' the number of methene groups, affects the fluorescence spectral characteristics.

Earlier studies of membrane potential with cyanine dyes employed bulk measurements and high dye concentrations. Under these conditions, cellular uptake of the cation during cell hyperpolarization results in *reduced* fluorescence of the suspension due to fluorescence quenching of the dye secondary to the formation of dye aggregates, and red-shifting of excitation and emission spectra (39). The use of more highly fluorescent dyes (see Chapter 2, *Table 3*) and flow cytometry allows adequate fluorescence signal detection of single cells with dye concentrations below 10^{-7} M. Under these conditions hyperpolarization is accompanied by increased cellular fluorescence and depolarization by decreased fluorescence. The lower dye concentrations also help to reduce the toxicity of these agents, as described below.

3.2.3 Oxonol dyes

The oxonol dyes are chemically unrelated to the cyanine dyes, but are similarly symmetrical, membrane-permeant molecules with a highly delocalized charge. This charge is negative, in contrast to the cyanine dyes, so that the changes in partitioning in response to the altered membrane potential are in the opposite direction to cyanine dyes. Cellular dye binding results in red-shifting of absorption and emission and/or an increase in quantum efficiency, which can be as high as 20-fold (40).

In earlier studies, oxonol dyes have not been used as extensively as the cyanine dyes for the measurement of membrane potential, in part because only a small proportion of dye is membrane-bound, and changes in the fluorescence of a suspension are small. For flow cytometry, however, the properties of the oxonol dyes are especially attractive (41):

- The analysis of cells takes place almost completely apart from dye in the external medium (increasing the signal-to-noise ratio).
- The lower proportion of bound dye increases the buffering of the system, minimizing fluctuations in external dye concentration.
- In addition, the negative charge of the dye forces exclusion from the highly negatively charged mitochondria, minimizing a complication encountered with cyanine dyes (see below).

3.3 Flow cytometric assay of membrane potential

3.3.1 General remarks

Flow cytometry was first demonstrated to be applicable to the analysis of membrane potential by Shapiro *et al.*, and the techniques used subsequently are fundamentally unaltered (42). As shown in *Protocol 3*, cells at 37°C are equilibrated with low concentrations of indicator dye, generally <50 nM cyanine dyes and <150 nM oxonol dyes (toxicity and other artefacts being reduced at lower concentrations of indicator). To keep the dye/cell ratio high and constant at these dye concentrations, the cell concentration should be $2-5 \times 10^5$/ml. After 5–15 min incubation the solution is introduced into the flow cytometer, the chamber (and in some cases the fluidics tubing) being warmed to 37°C, for the same reason previously described for calcium analysis.

Protocol 3

Typical preparation of cells for the analysis of membrane potential

Method

1. Incubate cells in medium at 4×10^5/ml with either 50 nM cyanine dye or 100 nM oxonol dye at 37°C for 10 min.

2. Prepare the flow cytometer by incubating the tubing with dye at the above concentration for 15 min at 37°C. If a previous sample has been analysed, wash by passing some of the dye solution forward through the tubing, rather than backflushing with saline.

3. Analyse unwashed cells in dye solution at 37°C.[a]

[a] When examining multiple aliquots of cells, the most satisfactory results will be obtained by analysis at a constant time after addition of the dye.

Various indicator dyes permit a range of choices in excitation and emission wavelength combination (Chapter 2, *Table 3*); however, for single-parameter measurements, excitation at 488 nm using DiOC$_5$(3) or di-BA-C$_4$(3) is most common. Adsorption of the dye on to the sample tubing may reach equilibration very slowly, and can result in slowly rising, baseline cellular fluorescence. This may be minimized by pre-treating the tubing with the dye solution without cells for 10 min or longer before an analysis. Alternatively, measurements can be made at a constant time after addition of the cell suspension to the flow cytometer. Forward-angle scatter is usually used to facilitate the discrimination of dead cells and cell aggregates; the former are obviously depolarized and will otherwise confuse the analysis. Analysis can also be gated on light scatter to examine only cells of a uniform size; unlike a ratio measurement, large cells, which bind more dye, will appear more fluorescent, independent of their membrane potential. As with calcium analysis, the resulting fluorescence profiles can be analysed and displayed as histograms, bivariate plots of fluorescence vs. time, or as mean fluorescence vs. time.

3.3.2 Calibration

Because the fluorescence changes of indicator dyes for a given membrane potential change are highly dependent upon the dye concentration, dye/cell ratio, and cell type utilized, calibration must ordinarily be established for each experiment. The most commonly used technique was originally described by Hoffmann and Laris (38) and employs the K^+ ionophore valinomycin to increase membrane K^+ permeability in the presence of various external K^+ concentrations. To find the membrane potential by the 'null point' method, the concentration of external K^+ is determined at which no change in the membrane potential-sensitive signal takes place. The membrane potential at this null point can then be calculated from the Nernst equation (see Section 3.2.1). A modification of this procedure is to add valinomycin to low $[K^+]$ medium, and then elevate the $[K^+]$ of the medium using concentrated KCl (40).

The oxonol dyes form complexes with valinomycin and this precludes the use of this ionophore for calibration. An alternative approach has been to take advantage of a calcium ionophore-induced increase in conductance of a Ca^{2+}-sensitive K^+ channel, establishing the membrane potential from the K^+ equilibrium potential (given by the Nernst equation), as above. In this manner, the oxonol fluorescence of A23187-treated T cells was found to vary linearly with the log of the external $[K^+]$ over a broad range (40).

3.3.3 Multiparameter analyses with membrane potential

Given the broad range of excitation and emission characteristics available with membrane potential-sensitive probes, it is perhaps not surprising that a variety of multiparameter assays are possible. As with analysis of indo-1, simultaneous membrane potential measurement and discrimination of monoclonal antibody fluorescence should be a valuable approach in elucidating subset-specific differences in cell response. One strategy for such analyses is based on the use of 488 nm excited membrane-potential probes (($DiOC_5(3)$, $DiOC_6(3)$, or $DiBaC_4(3)$)) and a red-excited antibody label such as PE–cyanine 5 or PE–Texas Red (see Chapter 2).

Witkowski and Micklem (43) used dye-laser excitation of the Texas Red to allow the estimation of membrane potential with $DiOC_6(3)$ in Lyt-2+ and Lyt-2– murine T-lymphocyte subsets and B cells. Chused and co-workers (41) used FITC antibodies and krypton laser excitation of $DiSBaC_2(3)$ to distinguish splenic T cells from B cells during analysis.

Indo-1 and membrane potential can be determined simultaneously using separate lasers for the UV excitation of indo-1, and visible excitation of any of the membrane-sensitive indicators. Lazzari *et al.* (44) analysed $DiOC_5(3)$ and indo-1 simultaneously, showing that the membrane potential changes in neutrophils stimulated with the oligopeptide chemoattractant f-MLP occurred just as rapidly as changes in $[Ca^{2+}]_i$.

Important insights can be obtained by combining other studies of cell function with membrane potential analysis. Seligmann *et al.* (45) analysed the binding of fluoresceinated f-MLPL oligopeptide simultaneously with $DiIC_5(3)$ fluorescence. These studies elegantly demonstrated that heterogeneity in membrane potential after oligopeptide exposure was related to differential binding of the chemoattractant. Simultaneous studies of forward and 90° scatter demonstrated that heterogeneity in $DiOC_5(3)$ fluorescence was not explained by heterogeneity in cell size. Vander Heiden and co-workers used rhodamine-123 to demonstrate that anti-apoptotic gene products can affect mitochondrial membrane potential, and concluded that Bcl-X promotes cell survival by regulating the electrical and osmotic homeostasis of mitochondria (46).

3.3.4 Limitations affecting the membrane potential assay

Membrane potential analysis was the first assay of signal transduction applied to flow cytometry. Unfortunately, this area is now lagging due to the lack of a suitable indicator for ratiometric analysis. When heterogeneous populations of cells are studied, this leads to a lower sensitivity of the assay to physiological cell signals than is encountered with calcium and pH assays that have 'modern' probes. However, a method using the ratio of oxonol fluorescence to fluorescence scatter signals has been developed that circumvents, in part, some of the limitations due to the use of probes without intrinsic ratiometric properties (47).

The cyanine dyes are recognized to have a variety of toxic and inhibitory effects, as well as certain limitations inherent in the use of cationic probes (reviewed by Chused *et al.* (41)). These dyes act to uncouple oxidative phosphorylation, deplete cellular ATP, block Ca^{2+}- dependent K^+ conductances, and cause depolarization of the resting membrane potential of lymphocytes and neutrophils. This toxicity can be reduced by using lower dye concentrations, but may not be completely eliminated in some cells types even in the nanomolar range.

Because mitochondrial potentials are highly negative, cyanine dye association with mitochondria can be a substantial component of the total fluorescence signal. Elaborate treatments to eliminate the mitochondrial potential by selective poisoning have been described (48); however, these may also affect the plasma membrane potential. A high ratio of dye/cells has been reported to decrease the proportion of cellular dye that is associated with mitochondria (40). The complications from mitochondrial dye uptake are less, of course, in mitochondria-poor polymorphonuclear leucocytes.

Finally, a suspension of cells in a cyanine dye may be poorly buffered, in that the major fraction of dye may be cell-associated. Extrusion of dye from a sub-population of cells may thus increase the available extracellular dye, which may then be taken up by another cell population.

As described previously, most of the above toxic limitations do not apply to the anionic oxonol dyes. The smaller fraction of dye associated with the cell results in reduced cellular fluorescence, but this has not been a problem with flow cytometry, and it results in better dye buffering. The negative charge of the oxonol dyes tend to exclude them from mitochondria, which reduces both toxicity and this component of the total fluorescence signal.

4 Measurement of cytoplasmic pH

4.1 Role of intracellular pH in cellular homeostasis

Intracellular pH (pH_i) is closely regulated by various mechanisms that, in mammalian cells, is primarily accomplished by $(Na^+)_o/(H^+)_i$ exchange and $(Cl^-)_o/(HCO_3^-)_i$ exchange. The Na^+/H^+ antiport is an electroneutral transport system that appears to have a major role in the regulation of pH_i. The antiport is stimulated by a variety of growth factors and is regulated by serine and tyrosine kinases and phosphatases in various cells. One of the earliest responses to the addition of mitogens is an increase in pH_i. An alkaline shift of ~0.25 units that peaks during S phase and mitosis occurs in cells (49). Other cellular responses are however, accompanied by a net cellular acidification, as, for example, when natural killer cells encounter target cells (50). The pH_i of mammalian cells is ~7.2 and appears to be closely regulated, as the coefficient of variation measured by flow cytometry is <5% (51).

A central question remains as to the potential role that pH_i might have in cellular signalling. Perhaps the clearest evidence for pH in cellular signalling derives from experiments in sea-urchin eggs where fertilization by sperm causes a transient rise in free calcium concentration followed by a prolonged rise in pH_i. The calcium signal can be bypassed by various parthenogenetic agents that directly elevate pH_i. A number of reports purporting to show a role for pH_i in cellular activation are derived from measurements made in bicarbonate-free medium. Thus, non-physiological (HCO_3-free) buffers are used because changes in pH_i are often only observed under bicarbonate-free conditions, presumably because changes in pH_i would be otherwise buffered by Cl^-/HCO_3^- exchange. Another concern is that reports attributing physiological effects to alterations of pH_i are often derived from cell lines and not normal cells. For example, the magnitude of change in pH_i after stimulation by serum of transformed fibroblasts is approximately twice that of normal cells (52).

4.2 pH probes

4.2.1 Fluorescein-based probes

There are several probes currently available for the measurement of pH_i using flow cytometry (see *Table 4*). Until recently, the most commonly used probes were modifications of fluorescein, fluorescein diacetate being the first-generation pH probe. Loading is accomplished by incubating cells with fluorescein diacetate; as with indo-1, the ester is hydrolysed and fluorescein is trapped within the cell. The use of fluorescein is limited, however, by the rapid leakage that occurs within minutes after loading cells. Carboxyfluorescein diacetate (CFDA) is similar to fluorescein diacetate except that it is more highly charged and, therefore, remains intracellularly trapped for longer. The loading of CFDA is facilitated by acidification of the medium, followed by resuspension of the cells in a neutral buffer after loading (53).

The most commonly used probe is 2',7'-bis-carboxyethyl-5(6)- carboxyfluorescein (BCECF). BCECF is loaded into cells using the acetoxymethyl ester; after hydrolysis, it has a negative charge of −4 or −5 and therefore leaks more slowly than CFDA. The pK_a of BCECF is 6.98, which is useful as it is near the pH_i of resting cells. There are pH-dependent shifts in the excitation wavelength for both CFDA and BCECF, and thus it is possible to use the ratio of fluorescence signals to correct for differences in loading and cell size. With BCECF, fluorescence excited at 450 nm is pH-independent, while fluorescence at 500 nm is pH-dependent.

The principal limitation with the use of fluorescein derivatives is that they exhibit relatively modest pH-dependent changes in fluorescence. For example, when the pH is lowered from 7.5 to 6.5, the ratio of fluorescence intensities after excitation of FDA at 436 nm and 495 nm increases by only 1.45 to 1.55 times (54). Given a coefficient of variation of 5–10% for the distribution of pH_i in

Table 4 Fluorochromes for ratiometric determination of pH using flow cytometry

Probe	pK_a	Fluorescence (nm) Excitation	Emission	Reference
Fluorescein diacetate (FDA)	6.3	ratio 436/495	525	54
Carboxyfluorescin diacetate (COFDA)	6.4	ratio 441/488	535	42
bis-Carboxyethyl-carboxyfluorescein acetoxymethyl ester (BCECF AM)	6.98	ratio 439/490 488	535 ratio 520/620	62 55
Hydroxycoumarin (4-methylumbelliferone, 4-MU)	7.8	~350 ~350	ratio 430/470 ratio 450/560	51 55
Diacetoxy-dicyanobenzene (ADB); yields dicyanhydroquinone (DCH) after de-esterification	8.0	~350	ratio 425/540	51, 55
Carboxy SNARF-1, acetoxymethyl acetate	7.5	488 or 514	ratio 575/670	57

a cell population, and the fact that physiological changes in pH are only <0.25 pH units, fluorescein-based probes can be expected to detect only extreme changes in pH_i.

4.2.2 pH probes used with UV excitation

There are several pH probes that are UV-excited. Cells can be loaded with 1,4-diacetoxy-2,3-dicyanobenzene (ADB) which is hydrolysed and trapped intracellularly to yield 2,3-dicyanohydroquinone (DCH). DCH fluorescence is pH-dependent, while the esterified forms exhibit pH-independent fluorescence; therefore, care must exercised to ensure complete de-esterification of the ADB. A major advantage of DCH, unlike fluorescein derivatives, is that ratiometric pH determinations are possible using a single excitation source by measuring fluorescence emission at 429 nm and 477 nm. Gerson *et al.* (49) have described a ratiometric technique using 4-methylumbelliferone (7-hydroxy-4-methyl coumarin). Others have found that background fluorescence may cause major limitations with the use of this probe (55). These probes have recently been used to demonstrate that neutrophils from patients with sepsis syndrome have elevated pH_i (56).

4.2.3 SNARF pH indicators

Perhaps the most useful probes for pH_i measurement are a family of compounds termed SNARF (SemiNaphthoRhodaFluor) and SNAFL (SemiNaptho-Fluorescein) (57). These probes were developed by investigators at Molecular Probes and were the first pH probes to have fluorescence characteristics that provide truly useful ratioing properties. These compounds appear to offer a technical improvement over previously available pH probes, similar in magnitude to the advances in calcium measurement realized with the introduction of indo-1. SNARF-1 is the reagent most suitable for use in flow cytometry. In contrast to the previously used probes, SNARF-1 exhibits large changes in pH-

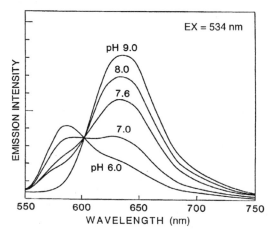

Figure 7 Fluorescence emission spectra of SNARF-1 as a function of pH. (Courtesy of R. Haugland, Molecular Probes.)

dependent fluorescence (see *Figure 7*). The acid form of SNARF-1 has absorption maxima at 518 and 548 nm, while the base form excites maximally at 574 nm. The emission of SNARF-1 in acid is maximal at 587 nm and the basic form emits maximally at 636 nm; there is an isosbestic point at 610 nm. In practice, either the 488 nm line or the 514 nm line of an argon-ion laser is suitable for excitation, while emission should be collected at both 575 nm and 670 nm and the ratio of these signals calculated. A protocol for the preparation of cells for use with SNARF-1 pH$_i$ measurement is shown below.

Protocol 4

Typical preparation of cells for analysis of pH$_i$ with SNARF-1

Reagents

- Hepes-buffered medium, pH 7.4, (with or without 2% serum)
- SNARF-1 acetoxymethyl ester (Molecular Probes)
- Buffer A: 135 mM KH$_2$PO$_4$, 20 mM NaCl, 1 mM MgCl, 1 mM CaCl$_2$, 10 mM glucose (see step 5)
- Buffer B: 135 mM K$_2$HPO$_4$, 20 mM NaCl, 1 mM MgCl, 1 mM CaCl$_2$, 10 mM glucose (see step 5)
- Nigericin (Sigma)

Method

1. Suspend the cells at ~10 × 10^6/ml in Hepes-buffered medium (with or without 2% serum).

2. Add 3–10 µM SNARF-1 acetoxymethyl ester and incubate at 37°C for 30 min. Determine the optimal concentration for each cell type.

3. Centrifuge the cells (100 g, 10 min, 22°C) and resuspend in fresh medium. Note that the cell pellet will appear pink on visual inspection. Store the cells at 22°C in the dark until analysis.[a]

4. Warm an aliquot of cells to 37°C for at least 5 min and analyse on a flow cytometer equipped with a temperature-control device. Excite fluorescence with either the 488 nm or the 514 nm line of an argon-ion laser and collect fluorescence at 575 nm and 670 nm. Note that the ratio of 670 nm/575 nm fluorescence intensity is proportional to increasing pH (see *Figure 8*).

5. For calibration of SNARF-1 fluorescence ratios to pH, prepare Buffer A and Buffer B. Mix Buffers A and B in various ratios to obtain a series of buffers that have a pH between 6 and 8, and an otherwise similar ionic constitution. Add 2 µg/ml of nigericin to the buffers. Suspend the cells (~5 × 10^5/ml) in the calibration buffers, equilibrate for 5 min at 37°C, and measure SNARF-1 fluorescence ratio (see *Figure 8* for example).

[a] Medium that is nominally HCO$_3^-$ free is often used in order to prevent HCO$_3^-$/Cl$^-$ exchange in cells.

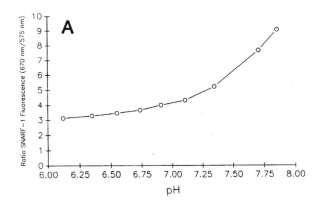

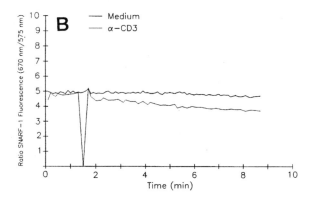

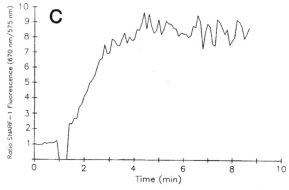

Figure 8 Calibration and analysis of cells with SNARF-1. The Jurkat cell line was loaded with the acetoxymethylester of SNARF-1 (see *Table 13.4*). (A) Cells were equilibrated at 37°C in buffers containing 135 mM K$^+$ at varying pH in the presence of nigericin. Fluorescence was excited by the 514 nm line of an argon-ion laser and the ratio of fluorescence emitted at 670 nm and 575 nm displayed. (B) Effect of stimulation of cells with anti-CD3 monoclonal antibody G19-4 or medium. A shift in the SNARF-1 fluorescence ratio corresponding to a cellular acidification of ~0.2 pH units occurs. (C) Ability of SNARF-1 to measure cellular alkalinization. Cells were suspended in a high potassium buffer (pH 8.98) and nigericin (2 μg/ml), a K$^+$/H$^+$ exchanger, was added during the gap in analysis.

4.2.4 Applications of SNARF-1

SNARF-1 has been used to study pH homeostasis in human keratinocytes (58). Simultaneous measurements of calcium and pH from conjugates of killer cells and target cells have been performed using fluo-3 and SNARF-1 (59). Recently, in studies of apoptosis in human T-cells induced by methyl mercury, a temporal relationship between mitochondrial dysfunction and loss of reductive reserve was demonstrated using a flow cytometric SNARF-1 assay (60). The toxin induced a decrement in pH_i of 0.5, and the authors concluded that the target organelle for MeHgCl toxicity is the mitochondrion and that induction of oxidative stress leads to activation of death-signalling pathways.

4.3 pH calibration

For each experiment, it is necessary to construct a calibration curve so that one may determine the fluorescence channel number as a function of pH_i. Cells are suspended in a series of high potassium buffers of different pH. The cells should be loaded with the pH probe and treated with the proton ionophore nigericin, added in order to equalize pH_i and to buffer pH (53). Nigericin produces the equilibrium $[K^+]_i/[K^+]_o = [H^+]_i/[H^+]_o$. Thus, in the presence of nigericin, pH_i can be calculated with the knowledge of $[K^+]_i$, $[K^+]_o$, and the buffer pH. *Figure 8A* illustrates such a calibration curve, obtained using the dye SNARF-1. This curve was acquired over a pH range encompassing that compatible with cell viability; much larger changes in SNARF-1 fluorescence ratio occur in medium that is more alkaline. *Figure 8B* demonstrates a pH_i shift in Jurkat cells following anti-CD3/T-cell receptor stimulation which produced an intracellular acidification in Jurkat cells. Human T cells appear to differ from rodent T cells that exhibit intracellular alkalinization after mitogenic stimulation (61). In *Figure 8C*, a pronounced shift in the SNARF-1 fluorescence ratio is shown after treatment of cells suspended in a high potassium buffer, pH 8.98, with nigericin. This result shows that it may be difficult to keep the ratio shifts on a linear scale if one examines large, non-physiological pH shifts with SNARF-1. By inspecting the data shown in *Figure 8A* and *C*, one can appreciate that the effective K_d of SNARF-1 is above that of the intracellular pH of resting cells, and that larger ratio shifts with SNARF-1 will occur upon cellular alkalinization than upon acidification.

Acknowledgements

This work was supported, in part, by National Institutes of Health Grant AG01751.

References

1. Bootman, M. D., Berridge, M. J., and Lipp P. (1997). *Cell*, **91**, 367.
2. Tsien, R. Y., Pozzan, T., and Rink, T. J. (1982). *J. Cell. Biol.*, **94**, 325.
3. Tsien, R. Y. (1981). *Nature*, **290**, 527.
4. Grynkiewicz, G., Poenie, M., and Tsien, R. Y. (1985). *J. Biol. Chem.*, **260**, 3440.

232

5. Haugland, R. P. and Johnson, I. D. (1999). In *Fluorescent and luminescent probes for biological activity* (2nd edn) (ed. W. T. Mason), p. 40. Academic Press, San Diego, CA.

6. Eberhard, M. and Erne, P. (1991). *Biochem Biophys Res Comm.*, **180**, 209.

7. Rabinovitch, P. S., June, C. H., Grossmann, A., and Ledbetter, J. A. (1986). *J. Immunol.*, **137**, 952.

8. Chused, T. M., Wilson, H. A., Greenblatt, D., Ishida, Y., Edison, L. J., Tsien, R. Y., and Finkelman, F. D. (1987). *Cytometry*, **8**, 396.

9. Zhao, M., Hollingworth, S., and Baylor, S. M. (1997). *Biophys. J.*, **72**, 2736.

10. Goller, B. and Kubbies, M. (1992). *J. Histochem. Cytochem.*, **40**, 451.

11. Kachel, V., Kempski, O., Peters, J., and Schodel, F. (1990). *Cytometry*, **11**, 913.

12. Moisescu, D. G. and Pusch, H. (1975). *Pflugers Archiv.*, **355**, R122.

13. Alberola-Ila, J., Takaki, S., Kerner, J. D., and Perlmutter, R. M. (1997). *Annu. Rev. Immunol.*, **15,** 125.

14. Brunkhorst, B. A., Lazzari, K. G., Strohmeier, G., Weil, G., and Simons, E. R. (1991). *J. Biol. Chem.*, **266**, 13035.

15. Haugland, R. P. (1992). In *Handbook of fluorescent probes and research chemicals* (5th edn) (ed. K. D. Larison), p. 117. Molecular Probes, Inc., Eugene, OR.

16. Minta, A., Kao, J. P., and Tsien, R. Y. (1989). *J. Biol. Chem.*, **264**, 8171.

17. Novak, E. J. and Rabinovitch, P. S. (1994). *Cytometry*, **17**, 135.

18. Ledbetter, J. A., June, C. H., Rabinovitch, P. S., Grossmann, A., Tsu, T. T., and Imboden, J. B. (1988). *Eur. J. Immunol.*, **18**, 525.

19. Schwinzer, R., Sommermeyer, H., Schlitt, H. J., Schmidt, R. E., and Wonigeit, K. (1991). *Cell Immunol.*, **136**, 318.

20. June, C. H., Ledbetter, J. A., Rabinovitch, P. S., Martin, P. J., Beatty, P. G., and Hansen, J. A. (1986). *J. Clin. Invest.*, **77**, 1224.

21. Grossmann, A. and Rabinovitch, P. S. (1987). In *Clinical cytometry and histometry* (ed. G. Burger, J. S. Ploem, and K. Goerttler), p. 192. Academic Press, London.

22. Davies, T. A., Weil, G. J., and Simons, E. R. (1990). *J. Biol. Chem.*, **265**, 11522.

23. Van Graft, M., Kraan, Y. M., Segers, I. M., Radosevic, K., De Grooth, B. G., and Greve, J. (1993). *Cytometry*, **14**, 257.

24. Alexander, R. B., Bolton, E. S., Koenig, S., Jones, G. M., Topalian, S. L., June, C. H., and Rosenberg, S. A. (1992). *J. Immunol. Meth.*, **148**, 131.

25. Goldsmith, M. A. and Weiss, A. (1988). *Science*, **240**, 1029.

26. Liddle, R. A., Misukonis, M. A., Pacy, L., and Balber, A. E. (1992). *Proc. Natl Acad. Sci. USA*, **89**, 5147.

27. Dunne, J. F. (1991). *Cytometry*, **12**, 597.

28. Philosophe, B. and Miller, R. A. (1989). *Eur. J. Immunol.*, **19**, 695.

29. Yamada, H., June, C. H., Finkelman, F., Brunswick, M., Ring, M. S., Lees, A., and Mond, J. J. (1993). *J. Exp. Med.*, **177**, 1613.

30. Szollosi, J., Feuerstein, B. G., Hyun, W. C., Das, M. K., and Marton, L. J. (1991). *Cytometry*, **12**, 707.

31. Sei, Y., McIntyre, T., Skolnick, P., and Arora, P. K. (1992). *Life Sci.*, **50**, 419.

32. Bernstein, R. L., Hyun, W. C., Davis, J. H., Fulwyler, M. J., and Pershadsingh, H. A. (1989). *Cytometry*, **10**, 469.

33. Sarvazyan, N., Swift, L., and Martinez-Zaguilan, R. (1998). *Arch. Biochem. Biophys.*, **350**, 132.

34. Huang, C. M., Fleisher, T. A., and Elin, R. J. (1988). *Magnesium*, **7**, 219.

35. Chien, M. M. and Cambier, J. C. (1990). *J. Biol. Chem.*, **265**, 9201.

36. Zhao, M., Hollingworth, S., and Baylor, S. M. (1996). *Biophys. J.*, **70**, 896.

37. Rijkers, G. T. and Griffioen, A. W. (1993). *Biochem. J.*, **289**, 373.

38. Hoffmann, J. F. and Laris, P. C. (1974). *J. Physiol.*, **239**, 519.

39. Waggoner, A. S. (1979). *Annu. Rev. Biophys. Bioeng.*, **8**, 47.
40. Rink, T. J., Montecucco, C., Hesketh, T. R., and Tsien, R. Y. (1980). *Biochem. Biophys. Acta*, **595**, 15.
41. Chused, T. M., Wilson, H. A., Seligmann, B. E., and Tsien, R. (1986). In *Applications of fluorescence in the biomedical sciences* (ed. D. L. Taylor, A. S. Waggoner, F. Lanni, R. F. Murphy, and R. R. Birge), p. 531. Alan R. Liss, New York.
42. Shapiro, H. M. (1990). In *Methods in cell biology* Vol. 33 (ed. Z. Darnzynkiewicz and H. A. Crissman), p. 25. Academic Press, San Diego.
43. Witkowski, J. and Micklem, H. S. (1985). *Immunology*, **56**, 307.
44. Lazzari, K. G., Proto, P. J., and Simons, E. R. (1986). *J. Biol. Chem.*, **261**, 9710.
45. Seligmann, B., Chused, T. M., and Gallin, J. I. (1984). *J. Immunol.*, **133**, 2641.
46. Vander Heiden, H. M., Chandel, N. S., Williamson, E. K., Schumacker, P. T., and Thompson, C. B. (1997). *Cell*, **91**, 627.
47. Seamer, L. C. and Mandler, R. N. (1992). *Cytometry*, **13**, 545.
48. Philo, R. D. and Eddy, A. A. (1978). *Biochem. J.*, **174**, 801.
49. Gerson, D., Kiefer, H., and Eufe, W. (1982). *Science*, **216**, 1009.
50. Edwards, B. S., Hoffman, R. R., and Curry, M. S. (1993). *J. Immunol.*, **150**, 4766.
51. Cook, J. A. and Fox, M. H. (1988). *Cytometry*, **9**, 441.
52. Gillies, R. J., Cook, J., Fox, M. H., and Giuliano, K. A. (1987). *Am. J. Physiol.*, **253**, C121.
53. Thomas, J. A., Buchsbausm, R. N., Zimniak, A., and Racker, E. (1979). *Biochemistry*, **18**, 2210.
54. de Grooth, B. G., van Dam, M., Swart, N. C., Willemsen, A., and Greve, J. (1987). *Cytometry*, **8**, 445.
55. Musgrove, E., Rugg, C., and Hedley, D. (1986). *Cytometry*, **7**, 347.
56. Rothe, G., Kellermann, W., and Valet, G. (1990). *J. Lab. Clin. Med.*, **115**, 52.
57. Bright, G. R., Whitaker, J. E., Haugland, R. P., and Taylor, D. L. (1989). *J. Cell Physiol.*, **141**, 410.
58. van Erp, P. E., Jansen, M. J., de Jongh, G. J., Boezeman, J. B., and Schalkwijk, J. (1991). *Cytometry*, **12**, 127.
59. Van Graft, M., Kraan, Y. M., Segers, I. M., Radosevic, K., De Grooth, B. G., and Greve, J. (1993). *Cytometry*, **14**, 257.
60. Shenker, B. J., Guo, T. L., and Shapiro, I. M. (1999). *Toxicol. Appl. Pharmacol.*, **157**, 23.
61. Hesketh, T. R., Moore, J. R., Morris, J. D., Taylor, M. V., Roger, J., Smith, G. A., and Metcalfe, J. C. (1985). *Nature*, **313**, 481.
62. Paradiso, A. M., Tsien, R. Y., and Machen, T. E. (1987). *Nature*, **325**, 447.

Chapter 14

Flow cytometry in the study of apoptosis

Michael G. Ormerod* (with contribution to Section 4 by Martin Poot[†])

*34, Wray Park Rd, Reigate, Surrey RH2 0DE, UK
[†]Martin Poot, University of Washington, Department of Pathology Box 357705, Seattle WA 98195, USA

1 Introduction

Cells have an in-built programme which, when triggered, causes the death of the cell. The mec1hanism by which programmed cell death is executed is called apoptosis. Apoptosis is involved in many normal biological processes, such as embryonic and T-cell development, metamorphosis, and hormone-dependent atrophy; it can also be induced by a variety of cytotoxic processes. There is now a huge literature on apoptosis; I have referenced some of the key reviews in a recent article (1).

Flow cytometry has found widespread application in the study of apoptosis; the literature is reviewed in references 1–3. In this chapter, I describe some of the methods used. Suitable methods have also been described by other authors (4–6).

The listed data files from which the figures in this chapter were drawn are available from the author on CD-ROM.

2 Morphological and biochemical changes

Apoptosis is characterized morphologically by condensation of nuclear chromatin, compaction of cytoplasmic organelles, cell shrinkage, collapse of the mitochondrial membrane potential, and changes at the cell surface. Apoptotic cells *in vivo* are rapidly phagocytosed, whereas rupture of the plasma membrane occurs only at a late stage *in vitro*. Usually apoptosis is accompanied by fragmentation of DNA into oligonucleosomal fragments with lengths that are multiples of 180–200 bp. In contrast, the hallmark of necrosis is uncontrolled swelling followed by rupture of the plasma membrane. DNA degradation tends to be non-specific.

Morphological changes in the nucleus can be observed by fluorescence microscopy. Incubate cells for 20 min at 37°C either with Acridine Orange

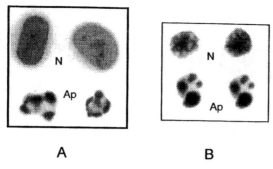

A   B

Figure 1 Nuclei from normal (N) and apoptotic (Ap) cells recorded by fluorescence microscopy. (A) Human ovarian carcinoma cell line incubated for 24 h with a thymidylate synthase inhibitor. Unfixed cells stained with Hoechst 33342. (Cells prepared by Julie M. Heward, Institute of Cancer Research, Sutton.) (B) Murine haematopoietic cell line, BAF3, 16 h after withdrawal of the growth factor, IL-3. Fixed cells stained with PI. (Cells prepared by Simone Detre, Royal Marsden Hospital, London.)

(20 μg/ml) or Hoechst 33342 (10 μg/ml). Alternatively, fixed cells can be stained with PI. The apoptotic cells are quite distinctive (see *Figure 1*). When embarking on a new study, the presence of apoptotic cells should always be confirmed by microscopy. Flow cytometry on its own should not be relied on to identify, without doubt, apoptotic cells. Most assays are capable of giving false-positive and false-negative results.

4 Measurement of DNA degradation

Two widely used methods for detecting and counting apoptotic cells are based on the extensive degradation of DNA during the later stages of apoptosis. If cells are fixed in ethanol and subsequently rehydrated, some of the lower molecular weight DNA leaches out, lowering the DNA content. These cells can be observed as a hypodiploid or 'sub-G_1' peak in a DNA histogram. Alternatively, the ends of the broken strands of DNA can be labelled enzymatically, called the ISEL (*in-situ* end-labelling) or TUNEL (Tdt-mediated, dUTP nick end-labelling) method. To apply this method, the cells are fixed in paraformaldehyde to cross-link the small fragments of DNA into the cell before fixation in ethanol.

In some cultured cell lines, one of the key components involved in the DNA degradation is absent or defective, in which case these methods cannot be applied (7).

3.1 Detecting apoptotic cells by measuring a DNA histogram

Use *Protocol 1* in Chapter 6 to prepare fixed cells for analysis of a DNA histogram. Leave the cells overnight at 4°C. This step gives time for the smaller pieces of DNA to diffuse out of the cell and gives a more reproducible separation between the G_1 and sub-G_1 peak. A typical result is shown in *Figure 2*. The apoptotic peak may be contaminated by necrotic cells and debris. Gating on light scatter may improve the resolution (see *Figure 2*).

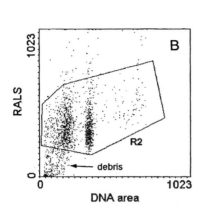
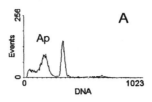
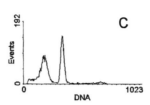

Figure 2 DNA histogram showing a 'sub-G_1' peak from apoptotic cells. Apoptosis was induced in a murine haematopoietic cell line (BAF3) by the withdrawal of the growth factor, interleukin-3 (IL-3). Murine haematopoietic cell line, BAF3, 16 h after withdrawal of the growth factor, IL-3. Fixed cells stained with PI. (A) DNA histogram gated on a cytogram of DNA peak vs. DNA area to exclude clumps. 'Ap' marks the position of the 'sub-G_1' peak. (C) The same DNA histogram after gating on a cytogram (B) of RALS vs. DNA to exclude debris. (Data recorded on a Coulter Elite ESP.)

The amount of DNA extracted from the cells, and hence the position of the apoptotic peak in the DNA histogram, depends on the type of cell under study. Sometimes the apoptotic cells may fail to give a 'sub-G_1' peak, although an inspection of a cytogram of DNA peak vs. DNA area may reveal the apoptotic cells (see *Figure 3*). In such cases, after ethanol fixation, the cells should be resuspended in a phosphate–citrate buffer (0.2 M Na_2HPO_4, 4 mM citric acid, pH 7.8), a treatment that will extract more DNA from the apoptotic cells (8).

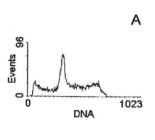
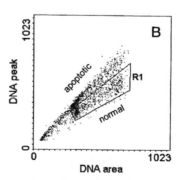

Figure 3 A sub-line of a human lymphoblastoid cell line, W1L2, incubated for 18 h with 2 mM KCN—a treatment that induces apoptosis in these cells. Cells were fixed and stained with PI. The DNA histogram (A) did not show a 'sub-G_1' peak. However, a cytogram of DNA peak vs. DNA area revealed two populations of cells. Cell sorting and microscopic examination confirmed the designation shown on the figure. A region, R1, has been drawn around the normal cells. (Data recorded on a Coulter Elite ESP.)

237

Care should always be taken in interpreting data from a DNA histogram. Apoptotic cells should be observed as a distinct peak. Necrotic cells, whose DNA is degraded randomly, will have a reduced DNA content and will be distributed across the same region of the histogram. If the instrument has been set to trigger on DNA/dye fluorescence (as is usually done when measuring a DNA histogram), an artificial 'sub-G_1' peak may be observed.

A linear amplifier should always be used to record the DNA histogram. There are examples in the literature of the use of a logarithmic amplifier. Not infrequently, the so-called sub-G_1 peak will be found to contain particles whose DNA content is a few per cent of that of cells in G_1. One has to question whether particles containing such small amounts of DNA should be counted as apoptotic cells.

3.2 The TUNEL assay

Using the enzyme, terminal deoxynucleotidyl transferase (Tdt), the broken ends of DNA can be labelled:

directly using:

- fluorescein–deoxyuridine triphosphate (dUTP),

or indirectly using:

- biotin–dUTP followed by fluorescein–streptavidin,
- digoxigenin–dUTP followed by fluorescein–anti-digoxygenin, or
- BrdUTP followed by fluorescein–anti-BrdUrd.

The cells should be fixed in ice-cold 1% paraformaldehyde for 15 min followed by 70% ethanol. The paraformaldehyde fixation cross-links the DNA into the cell and prevents the low molecular weight DNA from being extracted. After fixation, the cells can be stored indefinitely at –20 °C. When the DNA strand-breaks have been labelled, the cells are counterstained with PI so that the position in the cell cycle from which the cells committed apoptosis can be observed.

One can purchase all the necessary reagents separately and label the cells as in *Protocol 1*, which describes the BrdUTP method. I would recommend using one of the commercial kits supplied by a wide variety of companies. Suitable protocols are supplied with the kits. A typical result is shown in *Figure 4*.

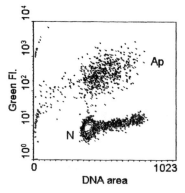

Figure 4 TUNEL assay. HL60 cells incubated with 10 μM camptothecin for 4 h. The cells were stained by Karen Holdaway (Beckman Coulter, Australia) according to *Protocol 1* using a kit supplied by Phoenix Flow Systems. A contour plot of green fluorescence (DNA strand-breaks) vs. DNA area. The apoptotic (Ap) and normal (N) cells are marked. Note that the cells had become apoptotic from the S phase of the cell cycle. (Data recorded on a Coulter XL.)

Labels other than fluorescein can be used and may be advantageous if the TUNEL assay is to be combined with an immunochemical stain.

Protocol 1

TUNEL method for labelling apoptotic cells

Reagents

- Phosphate-buffered saline (PBS): 1.06 g Na_2HPO_4, 0.39 g NaH_2PO_4, 8.8 g NaCl in distilled water. Adjust the pH to 7.2 and make up to 1 litre.

- 1% paraformaldehyde (Sigma) in PBS. Make up by gentle warming. Adjust the pH to 7.0. Either make up fresh or aliquot and store frozen.

- 1 mg/ml ribonuclease (RNase; Sigma) in distilled water. Make up fresh just before use.

- Tris-buffered saline (TBS): 150 mM NaCl in 50 mM Tris–HCl buffer pH 7.6. Dissolve 6.06 g Tris (base), 1.39 g Tris–HCl, 8.8 g NaCl in distilled water and make up to 1 litre.

- Reaction buffer (5 ×): 1 M Na cacodylate, 125 mM Tris–HCl pH 6.6, 1.25 mg/ml bovine serum albumin (BSA) (Cohn Fraction V, Sigma). Dissolve 1.96 g Tris–HCl, 16 g Na cacodylate, and 0.125 g BSA in distilled water and make up to 100 ml. Store at 4°C.

- Enzyme mixture: 20 U terminal deoxynucleotidyl transferase (Tdt) (Pharmacia), 2 μi 2 mM BrdUTP (Sigma), 5 μl 25 mM $CoCl_2$, 10 μl reaction buffer, and 35 μl distilled water. Make up immediately before use, for each sample to be analysed.

- Rinsing buffer: PBS, 0.2% Triton X-100, 1% BSA. Add 100 μl Triton X-100 to 100 ml PBS, 1% BSA; warm to 37°C to dissolve the Triton. Store at 4°C.

- Incubation buffer: 0.6 M NaCl, 60 mM Na citrate, 0.2% Triton X-100, 1% BSA. Dissolve 3.5 g NaCl, 1.76 g Na citrate, 1% BSA, and 100 μl Triton X-100 in 100 ml distilled water. Warm to 37°C. Store at 4°C.

- 1 mg/ml propidium iodide (PI; Molecular Probes) in 100 ml distilled water. Store at 4°C.

- Anti-BrdUrd–fluorescein (Becton Dickinson or Harlan Sera-Lab)

Method

1. Make a suspension of single cells to give a total of about 2×10^6 cells.

2. Centrifuge at 300 g for 5 min at room temperature. Discard the supernatant and resuspend the pellet thoroughly in 100 μl PBS.

3. Add 1 ml of ice-cold 1% paraformaldehyde solution. Leave on ice for 15 min.

4. Centrifuge as in step 2. Resuspend the pellet in ice-cold PBS. Centrifuge as before. Resuspend thoroughly in 200 μl of PBS.

5. Add, vigorously, 2 ml of ice-cold 70% ethanol. Leave for at least 30 min on ice. The cells can be stored indefinitely at –20°C in this state.

6. Centrifuge the cells at 400 g for 5 min at room temperature, discard the supernatant, and resuspend the pellet in TBS. Centrifuge the cells at 300 g for 4 min, resuspend the cell pellet in 50 μl of the enzyme mixture. Incubate at 37°C for 1 h.

7. Centrifuge the cells at 300 g for 4 min and resuspend the cell pellet in rinsing

MICHAEL G. ORMEROD

Protocol 1 continued

buffer. Repeat this washing procedure and finally resuspend the cell pellet in 100 μl of the incubation buffer containing 1 mg/ml anti-BrdUrd–fluorescein. Incubate on ice for 30 min.

8. Wash the cells twice in rinsing buffer (as in step 7) and resuspend the final pellet in 430 μl of PBS. Add 50 μl of the RNase solution and 20 μl of the PI solution. Incubate for 30 min at 37°C.

9. Analyse on the flow cytometer using blue light (argon-ion laser at 488 nm) and measuring forward- and right-angle light scatter, green fluorescence and red fluorescence. If possible, use doublet discrimination by measuring area and either the peak or width of the red fluorescent signal. Measure green fluorescence using a logarithmic amplifier, red using a linear amplifier. Set the discriminator (threshold) on the red fluorescent signal. Using pulse-shape analysis of this signal (see Chapter 6, Section 3.2.2), set a region on the single cells; use this region as a gate to display a cytogram of green fluorescence (fluorescein; strand breaks) vs. red fluorescence (PI; DNA). The apoptotic cells are green-positive.

4 Mitochondrial membrane potential

A drop in the mitochondrial membrane potential (MMP) is an important feature of the apoptotic cascade (9). There are fluorescent dyes that are sequestered in the mitochondria; a drop in the MMP causes a drop in the amount of dye in the cell. The most commonly used dyes are rhodamine-123 and 3,3'-dihexyloxacarbocyanine—DiOC$_6$(3). *Protocol 2* describes the use of the latter with an example of the result shown in *Figure 5*.

Another dye that has been used is a cationic carbocyanine, JC-1. JC-1, which can be excited at 488 nm, as a monomer, fluoresces green; when it is accumu-

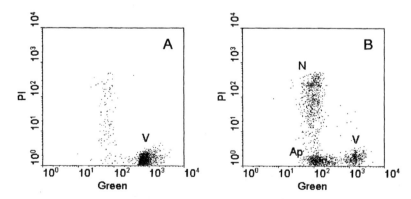

Figure 5 Assay for mitochondrial membrane potential using DiOC$_6$(3). A sub-line of W1L2 incubated for 18 h with 2 mM KCN. The cells were stained according to *Protocol 2* by Lorraine Skelton, Institute of Cancer Research, Sutton. The viable (V), apoptotic (Ap), and necrotic (N) cells are marked. The necrotic cells are clearly apoptotic cells undergoing secondary necrosis. (Data recorded on a Coulter Elite ESP.)

lated in mitochondria it aggregates and fluoresces red (see Chapter 2, *Table 3*). The green/red fluorescence ratio will decrease if the MMP drops (10).

Protocol 2

Following changes in the mitochondrial membrane potential

Reagents

- Stock solution of 1 mM DiOC$_6$(3) (Molecular Probes). Dissolve 5.73 mg in a small amount of dimethylsulfoxide (DMSO). Make up to 10 ml in water. Store frozen in 500 μl aliquots. Dilute to 5 μM (1:200) in water before use.

- 1 mg/ml PI (Molecular Probes)

Preliminary experiments

1. Treat cells with a suitable agent to give a mixture of normal and apoptotic cells (check microscopically).

2. Prepare a suspension of single cells in PBS at about 10^6 cells/ml. Split into four aliquots of 500 μl. Add DiOC$_6$(3) at 10, 30, 100, and 300 nM; (1, 3, 10, and 30 μl of 5 μM solution).

3. Incubate at 37 °C for 15 min. Add 2.5 μl PI solution (final concentration: 5 μg/ml).

4. Analyse, recording forward- and right-angle light scatter, log green and log red fluorescence.

5. Select the concentration of DiOC$_6$(3) that gave the best separation between normal (bright green, red-negative), apoptotic (weak green, red-negative), and necrotic cells (weak green, red-positive) (see *Figure 5*).

Subsequent experiments

1. Prepare suspensions of single cells in PBS at about 10^6 cells/ml from the control and test samples.

2. Add DiOC$_6$(3) at the concentration determined above.

3. Proceed as in steps 3 and 4 above.

The method described above assumes that the total mitochondrial mass is unchanged. However, the fluorescence obtained after staining with an MMP sensitive dye reflects the MMP per unit mitochondrion multiplied by the number of mitochondria per cell. Hence, a cell with more mitochondrial mass will be more fluorescent than a cell with less mitochondrial mass. It is possible to correct for possible differences in mitochondrial mass by counterstaining the cells with a mitochondria-specific dye that is not sensitive to the mitochondrial membrane potential. This dye has to emit at wavelengths that are different form those of the MMP-sensitive dye. By dividing the fluorescence intensity of the MMP-sensitive dye by the fluorescence intensity from the insensitive dye, the normalized MMP of each individual cell is obtained. A dye pair that has worked successfully in this way is MitoTracker Green FM (MTG; MMP-insensitive) and CMXRosamine (MitoTracker Red CMXRos; MMP-sensitive) (11).

Protocol 3

Measurement of the normalized mitochondrial membrane potential (supplied by Martin Poot)

Reagents

- 200 μM stock solutions of CMXRos (Molecular Probes) in DMSO[a]
- 200 μM stock solutions of MTG (Molecular Probes) in DMSO[a]

Method

1. Take 1 ml aliquots of cell suspensions and pre-warm to 37°C.

2. Add MTG and CMXRos to the cell suspensions to obtain a final concentration of 100 nM.[b]

3. Incubate cell suspensions for 30 minutes at 37°C in the dark. After incubation place the cell suspensions in a melting ice-bath.

4. Assay the cell suspensions immediately.[c]

[a] Stock solutions can be stored at –20°C in the dark for several weeks.

[b] Higher than 100 nM dye concentrations may result in cell staining that is not specific for mitochondria.

[c] Both dyes are excitable with the 488 nm line of an argon laser; MTG emits fluorescence around 530 nm, while CMXRos is best detected at wavelengths above 610 nm. To minimize possible interference from MTG fluorescence a 640 long-pass filter is preferred.

The ratio of CMXRos to MTG fluorescence intensity for each individual cell represents its normalized mitochondrial membrane potential. Since the MTG and the CMXRos dyes emit green and red fluorescence after excitation with 488 nm laser light, it is possible to combine this assay with quantification of NADH by UV-excited blue fluorescence (12). In this way, a combined assay for normalized mitochondrial membrane potential and normalized NADH content is obtained. Alternatively, cells can be co-stained with 10 μM of the Hoechst 33342 dye. This dye is taken up by all (viable and non-viable) cells, and after UV-excitation they emit blue fluorescence in proportion to cellular DNA content. Thus, an assay for cell-cycle stage distributions of 'normal' and apoptotic cells may be obtained.

5 Measurements of changes in the plasma membrane

Apoptosis induces a variety of changes in the plasma cell membrane, including changes in permeability and alterations in the membrane lipids. During apoptosis there is a major change in the membrane lipids in that phosphatidyl serine (PS) residues 'flip' from the internal to the external membrane. PS binds Annexin V and these changes can be observed by incubating unfixed cells with Annexin V (13). Permeability changes can be observed using one of several DNA-binding dyes, such as 7-aminoactinomycin D (7-AAD) (14), YO-PRO-1 (15), and Hoechst 33342 (16). These dyes are normally excluded by the plasma membrane but they can permeate apoptotic cells. Protocols are given below for the use of YO-PRO-1 and 7-AAD.

5.1 Binding of Annexin V to cells

Annexin V is a protein that has a high affinity for negatively charged phospholipids, such as PS, in the presence of Ca^{2+} ions. Unfixed cells are incubated with Annexin V conjugated to a suitable fluorochrome. Several manufacturers supply Annexin V conjugated to a variety of fluorochromes. It is customary to add PI to distinguish cells that have lost integrity of the plasma membrane, although, if Annexin V is to be combined with immunochemical labels, PI could be omitted.

Protocol 4

Detection of apoptotic cells through binding of Annexin V (13)

Reagents

- PBS, pH 7.3
- 10 mM Hepes/NaOH pH 7.4, 140 mM NaCl, 2.5 mM $CaCl_2$
- FITC–Annexin V
- PI

Method

1. Prepare a single cell suspension at $1\text{--}2.10^6$ cells/ml. Wash the cells in PBS.

2. Centrifuge and resuspend in Hepes/NaOH pH 7.4, 140 mM NaCl, 2.5 mM $CaCl_2$. Add FITC-Annexin V to a final concentration of 1 μg/ml.

3. Incubate for 10 min in the dark at room temperature.

4. Add PI to a final concentration of 2 μg/ml. Incubate for a further 5 min.

5. Analyse, recording right-angle and forward light scatter, log green (520 nm) and log red fluorescence (>650 nm). (Use the fluorescein filter set for green fluorescence and the deep red set (FL3) for red.)

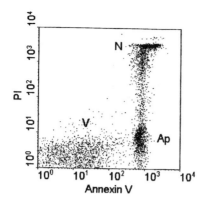

Figure 6 Annexin V assay. Human peripheral blood lymphocytes from a patient with chronic lymphocytic leukaemia were incubated for 48 h with 2 μg/ml chlorambucil. The cells were stained according to *Protocol 3*. The viable (V), apoptotic (Ap), and necrotic (N) cells are marked. (Data recorded on a Becton Dickinson FACScan and supplied by Chris Pepper, Department of Haematology, Llandough Hospital, Penarth, Wales.)

Figure 6 shows a typical result. A gate can be set on light scatter to exclude clumps of cells and debris, but be careful not to exclude any apoptotic cells which have changed light scatter compared to the viable cells.

5.2 Measuring changes in membrane permeability

Apoptotic cells are frequently more permeable than normal cells (17). The reason has not been elucidated and not all cells exhibit this property. Two protocols are given below. It may be necessary to vary the dye concentration and the incubation time to obtain the best separation between normal and apoptotic cells.

Protocol 5

Detection of apoptotic cells using YO-PRO-1 uptake (15)

Reagents
• YO-PRO-1 (Molecular Probes)   • PI

Method
1. Prepare a single-cell suspension at about 10^6 cells/ml.
2. Add YO-PRO-1 to 10 μM. Incubate at 37°C for 10 min.
3. Add PI to 5 μg/ml.
4. Analyse, recording right-angle and forward light scatter, log green (520 nm) and log red fluorescence (>600 nm). The apoptotic cells are bright green, red-negative; the necrotic/late stage apoptotic cells are bright green, red-positive; and the normal cells are weak green, red-negative (see *Figure 7*).

The advantage of the method using 7-AAD is that it can easily be combined with an immunochemical stain for cell-surface antigens (see *Protocol 6* and *Figure 8*).

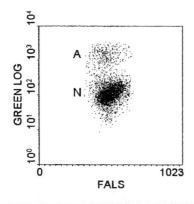

Figure 7 Enhanced uptake of the cyanine-based dye, YO-PRO-1, by apoptotic cells. A murine haematopoietic cell line, BAF3, incubated for 16 h without IL-3. Cells were stained according to *Protocol 5*. A cytogram is shown of green (YO-PRO-1) fluorescence vs. FALS after gating to select the PI-negative cells. A, apoptotic cells; N, normal cells. (Data recorded on a Coulter Elite ESP.)

Protocol 6

Detection of apoptotic cells using 7-AAD uptake (15)

Reagents

- Washing buffer: PBS plus 1% BSA, 0.1% Na azide
- 7-AAD (Molecular Probes)

Method

1. Prepare a single-cell suspension at about 10^6 cells/ml.

2. If required, stain the unfixed cells with antibodies labelled with fluorescein and PE (see Chapter 5).

3. Wash once in washing buffer, centrifuging the cells at 300 g for 5 min at room temperature. Resuspend the cells in 500 μl of the washing buffer containing 20 μg/ml 7-AAD.

4. Incubate on ice for 20 min.

5. Analyse, recording right-angle and forward light scatter, log green (520 nm) and log red fluorescence (>650 nm). (Use the fluorescein filter set for green fluorescence and the deep red set (FL3) for red.)

6 Other measurements

A variety of other measurements can be made to characterize apoptotic cells and to follow various events during the apoptotic cascade. Any protein involved in apoptosis can be measured, if a suitable antibody is available, using the methods described in Chapter 9.

If an appropriate substrate can be obtained, intracellular enzyme activity can be measured (Chapter 2, Section 5.7). In particular, the activity of caspase 3, a key enzyme involved in cell death, can be measured using the polypeptide, DEVD, linked to rhodamine-110.

A wide variety of changes to intracellular components can be followed by flow cytometry. These include glutathione (Chapter 15, Section 5.3), reactive

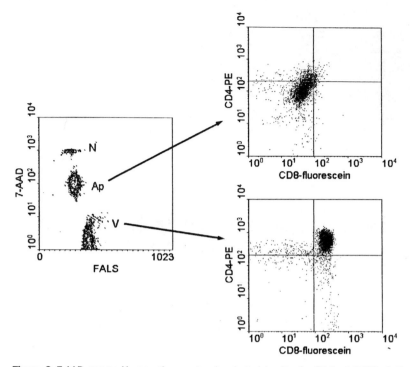

Figure 8 7-AAD assay. Human thymocytes incubated *in vitro* for 24 h at 37 °C. Cells were stained for the T-cell markers CD8 and CD4; and then with 7-AAD according to *Protocol 6*. The viable (V), apoptotic (Ap), and necrotic (N) cells are marked in the contour plot of 7-AAD fluorescence vs. forward-angle light scatter. The CD4/CD8 expression of the apoptotic and the viable cells are shown in cytograms A and B, respectively. Note that both CD4 and CD8 are down-regulated in the apoptotic cells. The quadrant gate has been set to show up the difference between the two types of cell; it was not set on the isotype control. (Data recorded on a Becton Dickinson FACScan and supplied by Ingrid Schmid, UCLA School of Medicine, Los Angeles, USA.)

oxidative species (Chapter 2, Section 5.7), calcium ions (Chapter 13, Section 1), and intracellular pH (Chapter 13, Section 3).

7 Quantification of apoptotic cells

The methods described above can be used to follow changes in apoptotic cells; they can also be used to count the percentage of apoptotic cells in a culture. For quantification of the cells, the simplest (and cheapest) method is the observation of a sub-G_1 peak in a DNA histogram. The sub-G_1 peak may be contaminated with fragmented material from necrotic cells or from cells damaged during sample preparation. The limit of sensitivity of the method is about 2% of the total cells present and, for this reason, the method is unsuitable for use with clinical samples.

The Annexin V method has the advantage of simplicity, although the reagents are more expensive. It is easy to combine Annexin V labelling with antibody labelling so that the immunophenotype of the apoptotic cells can be observed. The 7-AAD method also has this advantage and is cheap. Because it relies on a change in membrane permeability, it may not be universally applicable.

To date, the TUNEL assay has been used for clinical studies (for example, see ref. 18). It is probably the most sensitive method for samples that may contain debris and damaged cells. If small numbers of apoptotic cells are detected, it is helpful if the positive cells can be sorted for microscopic examination (19).

Acknowledgements

I thank Chris Pepper, Department of Haematology, Llandough Hospital, Penarth, Wales, and Ingrid Schmid, UCLA School of Medicine, Los Angeles, USA, for supplying me with the data used to construct *Figures 6 and 8*.

References

1. Ormerod, M. G. (1998). *Leukemia*, **12**, 1013.
2. Darzynkiewicz, Z., Bruno, S., Del Bino, G., Gorczyca, W., Hotz, M. A., Lassota, P., and Traganos, F. (1992). *Cytometry*, **13**, 795.
3. Darzynkiewicz, Z., Juan, G., Li, X, Gorczyca, W., Murakami, T., and Traganos, F. (1997). *Cytometry*, **27**, 1.
4. Darnzynkiewicz, Z., Li, X., and Gong, J. (1994). In *Methods in cell biology*, Vol. 41 (ed. Z. Darzynkiewicz, J. P. Robinson, and H. A. Crissman), p. 15. Academic Press, San Diego, CA.
5. Fraker, P. J., King, L. E., Lill-Elghanian, D., and Telford, W. G. (1995). In *Methods in cell biology*, Vol. 46 (ed. L. M. Schwartz and B. A. Osborne), p. 57. Academic Press, San Diego, CA.
6. Sherwood, S. W. and Schimke, R. T. (1995). In *Methods in cell biology*, Vol. 46 (ed. L. M. Schwartz and B. A. Osborne), p. 77. Academic Press, San Diego, CA.
7. Ormerod, M. G., O'Neill, C. F., Robertson, D., and Harrap, K. R. (1994). *Exp. Cell Res.*, **206**, 231.
8. Gong, J., Traganos, F., and Darzynkiewicz, Z. (1994). *Anal. Biochem.*, **218**, 14.
9. Zamzami, N., Marchetti, P., Castedo, M., Zanin, C., Vaysière, J-L., Petit, P. X., and Kroemer, G. (1995). *J. Exp. Med.*, **181**, 1661.
10. Petit, P. X., LeCoeur, H., Zorn, E., Dauguet, C., Mignotte, B., and Gougeon, M. L. (1995). *J. Cell Biol.*, **130**, 157.
11. Poot, M. and Pierce, R. H. (1999). *Cytometry*, **35**, 311.
12. Thorell, B. (1983). *Cytometry*, **4**, 61.
13. Van Engeland, M., Nieland, L. J. W., Ramaekerss, F. C. S., Schutte, B., and Reutelingsperger, C. P. M. (1998) *Cytometry*, **31**, 1.
14. Schmid, I., Uitenbogaart, C. H., Keld, B., and Giorgi, J. V. (1994). *J. Immun. Meth.*, **170**, 145.
15. Idziorek, T., Estaquier, J., De Bels, F., and Ameisen, J.-C. (1995). *J. Immun. Meth.*, **185**, 249.
16. Ormerod, M. G., Collins, M. K. L., Rodriguez-Tarduchy, G., and Robertson, D. (1992). *J. Immunol. Meth.*, **153**, 57.
17. Ormerod, M. G., Sun, X-M., Snowden, R. T., Davies, R., Fearnhead, and Cohen, G. M. (1993). *Cytometry*, **14**, 595.
18. Gorczyca, W., Bigman, K., Mittelman, A., Ahmed, T., Gong, J., Melamed, M. R., and Darzynkiewicz, Z. (1993). *Leukaemia*, **7**, 659.
19. Dowsett, M., Detre, S., Ormerod, M. G., Ellis, P. A., Mainwaring, P. N., Titley, J. C., and Smith, I. E. (1998). *Cytometry*, **32**, 291.

Chapter 15

Further applications to cell biology

Michael G. Ormerod

34, Wray Park Rd, Reigate, Surrey RH2 0DE, UK

1 Introduction

There are a multitude of applications for flow cytometry, the most important of which have been covered in earlier chapters. In this chapter, five further applications to cell biology are described. The first is a method for estimating cell viability; the second, a way of monitoring the effect of high electric fields on cells; the third, a method for measuring the production of oxidative species in cells; and the last section discusses two methods for characterizing drug resistance in cancer cells, including a method for measuring intracellular glutathione.

2 Estimating cell viability

2.1 Introduction

Traditionally, the number of live cells in a suspension is estimated by counting, microscopically, the cells which exclude an acidic dye, such as Trypan Blue. The method can be adapted for the flow cytometer by using propidium iodide (PI) which is excluded by viable cells and, when taken up by dead or dying cells, binds to nucleic acids and fluoresces red (see Chapters 2 and 14). The use of flow cytometry has the advantage that large numbers of cells can be counted quickly and that the determination of negative/positive cells is objective.

The resolution of the method can be improved by the additional use of fluorescein diacetate (FDA) or one of its derivatives. FDA, which is not fluorescent, is taken up by cells where it is converted to fluorescein by the intracellular esterases. Fluorescein is retained by the cell if the plasma membrane is intact.

2.2 Description of the method

The method is described in *Protocol 1* and a typical result shown in *Figure 1*. The cells are incubated in FDA and PI and should be kept on ice if any time elapses between incubation and analysis. This will slow down further hydrolysis of the FDA and leakage of fluorescein from the cells.

Protocol 1

Estimating cell viability

Reagents

- 100 μg/ml FDA (Sigma or Molecular Probes) in acetone. Dilute in PBS to give a working solution at 100 ng/ml.

- 100 μg/ml PI (Sigma or Molecular Probes) in PBS, pH 7.3

Method

1. Take 500 μl of a suspension of single cells at 10^6 cells/ml in any suitable buffer or culture medium.

2. Add 50 μl of the FDA solution and 50 μl of the PI solution. Stand at room temperature for 10 min.

3. Analyse, recording forward and orthogonal light scatter, red (>630 nm) and green (520 nm) fluorescence. Gate on light scatter to include single cells and to exclude clumps and debris. Display green vs. red fluorescence.[a]

[a]A cell whose plasma membrane is intact fluoresces green and not red.

Dead cells usually show decreased forward light scatter and slightly increased orthogonal scatter. The light-scatter gate should therefore be set generously to include all the dead and live cells but to exclude clumps of cells and debris, as far as possible.

In the cytogram of red vs. green fluorescence, the viable cells are counted as those that are positive for green and negative for red fluorescence. There may

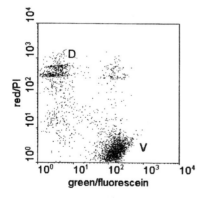

Figure 1 Viability assay of a human ovarian carcinoma cell line. The cells grow attached to the culture dish; the adherent cells were harvested using trypsin/EDTA and mixed with the non-adherent cells collected from the culture supernatant. The cells were then labelled in suspension with FDA and PI. The viable cells (V) are green-positive, red-negative; the dead cells (D), green-negative, red-positive. The green-positive, red-positive cells are probably clumps of cells that fell within the light-scatter gate. (The cells were prepared by Swee Sharp, Institute of Cancer Research, Sutton, and analysed by Jenny Titley on a Coulter Elite ESP.)

also be an intermediate population that shows both weak green and red fluorescence. These are probably dying cells, possibly apoptotic, and should not be counted in the viable population.

The method measures cells with an intact plasma membrane. The phrase 'viable cell' is slightly misleading since such cells may not be capable of division —for example, after a lethal dose of high-energy radiation.

3 Monitoring electropermeabilization of cells

3.1 Introduction

Electroporation is an efficient way of introducing foreign molecules, such as DNA, into cells. The cells are permeabilized at a low temperature by a high-voltage pulse and are then warmed to allow their membranes to reseal.

A variety of parameters, such as the medium used, the pulse voltage, and the length and number of pulses, require optimization. Cells also vary in their electrosensitivity. To establish the correct conditions, it is necessary to know whether holes have been punched in the plasma membrane, for how long they stay open, how efficiently they reseal, and the effects of this treatment on cell viability. The method described below has been developed to monitor electroporation in order to optimize these conditions (1).

3.2 Description of the method

This method is a variant of that used to estimate cell viability (see Section 2 above). PI is added immediately after electropermeabilization so that those cells whose membranes have been ruptured take up the dye and their nuclei fluoresce red. After the cells have been incubated in warm medium to allow their membranes to reseal, FDA is added so that the cells with intact membranes (either unpermeabilized or whose membranes have successfully resealed) fluoresce green. The method is outlined in *Figure 2*.

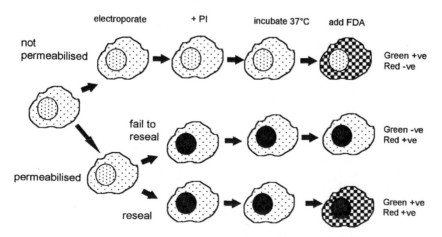

Figure 2 Method for measuring the effect of electroporation on cells by flow cytometry.

Protocol 2

Monitoring electroporation

Reagents

- Electroporation buffer: 0.3 M mannitol, 0.5 mM $CaCl_2$, 0.2 mM $MgCl_2$, 200 g/ml BSA (Cohn fraction V, Sigma) in distilled water
- Culture medium plus 10% fetal calf serum
- 100 µg/ml PI (Sigma or Molecular Probes) in PBS, pH 7.3
- 100 µg/ml FDA (Sigma or Molecular Probes) in acetone. Dilute in PBS to give a working solution at 100 ng/ml FDA

Method

1. Prepare a suspension of single cells at about 5×10^6 cells/ml in electroporation buffer. Cool on ice.

2. Cool the electroporation chamber in ice. Electroporate 200 µl of cells at 0 °C.

3. Add 20 µl of PI solution. Incubate at 0 °C for 10 min.

4. Dilute the cells into 1 ml culture medium plus 10% fetal calf serum at 37°C. Incubate for 10 min.

5. Wash the cells in PBS by centrifuging at 300g for 5 min and resuspend in 1 ml of PBS.

6. Add 20 µl of the FDA solution. Stand at room temperature for 10 min and then place in ice.

7. Analyse using an argon-ion laser tuned to 488 nm and recording orthogonal and forward light-scatter, red (>630 nm) and green (520 nm) fluorescence. Gate on the light scatter from single cells and display green vs. red fluorescence.

The details of the optimization of the parameters for electropermeabilization using a purpose-built, commercial apparatus (TA 750 from Krüss GmbH) have been given by O'Hare *et al.* (1). The flow cytometric method is described in *Protocol 2* and a typical result shown in *Figure 3*.

In the cytogram shown in *Figure 3*, three classes of cells are distinguished: fluorescein-positive, PI-negative (unpermeabilized); fluorescein-positive, PI-positive (permeabilized and resealed membranes); fluorescein-negative, PI-positive (permeabilized and failed to reseal). The unpermeabilized cells were present because a small number of cells were retained in the tip of the pipette and were not subjected to the electric field. They served as a useful inbuilt control.

The optimal voltage for electroporation depends on the type of cell being used. The first time a cell line is to be used for electrotransfection, the best DC voltage should be determined.

Electroporate the cells at a series of DC voltages and take aliquots for flow cytometry as described above. Set up cultures with second aliquots and assess their viability (see Section 2) after 18 h. Select the correct DC voltage by

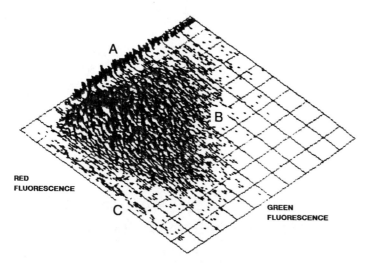

Figure 3 A cytogram of green vs. red fluorescence from cultured bovine endothelial cells, electroporated and then stained with PI and FDA. A, Unpermeabilized cells; B, permeabilized cells whose membranes have resealed; C, dead cells. (Data recorded on an Ortho Cytofluorograf 50H.)

comparing the survival curve with the flow cytometric data, selecting the best compromise between cell survival and permeabilization.

4 Measurement of oxidative burst (2, 3)

Phagocytes produce a variety of oxidative species when they are activated or when they ingest foreign particles. There are several probes that, when oxidized, produce fluorescent compounds and can be used to explore oxidative burst.

Compounds used to measure oxidative burst include dichlorodihydrofluor-

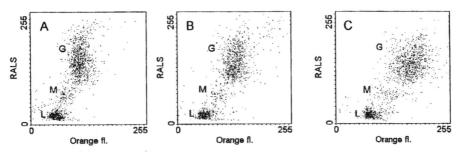

Figure 4 Oxidative burst in human leucocytes. Cells loaded with HE and incubated for 15 min with activating agents. (A) Unstimulated; (B) incubated with PMA; (C) incubated with 10 μM muramyl dipeptide (a bacterial cell-wall component). Granulocytes (G), monocytes (M), and lymphocytes (L) can be identified by their RALS. Activation of the cells caused an increase in orange fluorescence in the monocytes and granulocytes. (Data recorded by Richard Allman and Amanda J. Taylor, Velindre Hospital, Cardiff, Wales, on a Becton-Dickinson FACScan.)

escein (produces dichlorofluorescein); dihydrofluorescein (produces fluorescein); dihydrorhodamine-123 (produces rhodamine-123), and dihydroethidium (HE) (produces ethidium). Both the fluorescein-based compounds have to be loaded into the cells in the form of the diacetate ester (see Chapter 2, Section 5.1). HE has the advantage that that its product, ethidium, binds to DNA and therefore will not leak out of the cell during the course of the experiment (3).

There are always oxidative species being produced in all cells. Once the cells have been loaded with the probe, fluorescence will slowly increase with time. It is important that stimulated and unstimulated cells are always compared.

A method is given in *Protocol 3* and a typical result shown in *Figure 4*.

Protocol 3

Measurement of oxidative burst in human leucocytes

Reagents

- PBS gel stock: 0.2 M EDTA, 0.5 M dextrose, 10% gelatine in water. Heat water to 45–50 °C and add gelatine while mixing. Add EDTA and dextrose. Aliquot and store frozen.
- PBS gel: 1 ml gel stock in 100 ml PBS (warm to 45 °C before mixing) pH 7.4

- HE stock: 3 mM HE (Molecular Probes) in DSMO
- PMA stock: 1 mM PMA in DSMO. Aliquot and store frozen.
- PMA working solution (prepare daily). Dilute PMA stock 1:100 in PBS gel.

Method

1. Prepare a suspension of human peripheral blood leucocytes by lysing red cells (see Chapter 3, *Protocol 3*).

2. Suspend cells in 1 ml PBS gel at a concentration of about 10^6 cells/ml.

3. Add HE at a final concentration of 3 μM (1 μl stock solution in 1 ml cell suspension); incubate at 37 °C for 15 min.

4. Add PMA stock at 1:10 (final concentration, 1 μM); incubate at 37 °C for 15 min.

5. Place on ice and analyse without delay, measuring right-angle and forward light scatter and red fluorescence.

6. Display right-angle light scatter vs. red fluorescence.

5 Characterizing multi-drug resistance (MDR) in cancer cells

5.1 Introduction

Many malignant tumours develop resistance to a wide range of cytotoxic drugs. There are several causes of MDR. There are proteins in the plasma membrane which pump toxins and other chemicals out of cells. Overexpression of one of these pumps can prevent a drug accumulating within the cell.

One such pump, which has been well characterized, is a high molecular weight (170 kDa) glycoprotein (gp170) produced by the *MDR1* gene. Its over-expression can be detected by an antibody. However, the presence of the antigen does not indicate whether the pump is active or not. The antigenic epitope on gp170 can also be masked by sialic acid and it may be necessary to treat the cell with neuraminidase to obtain a true picture of gp170 expression. Furthermore, this protein is one of a family of related proteins, not all of which will be detected by a single antibody. A second type of pump, referred to as the MDR-associated protein (MRP) has been described. For these reasons, functional assays of the membrane pumps are often used.

Another cause of MDR is the overexpression of glutathione (GSH). Many cytotoxic drugs are oxidizing agents and will be rendered non-toxic by reaction with GSH.

5.2 Functional assay of MDR pumps (4)

The MDR pump can be inactivated by calcium channel blockers, such as verapamil and cyclosporin A (CyA). In the functional assay, cells are incubated with a fluorescent compound which is acted on by the gp170 pump (for example, rhodamine-123, Hoechst 33342, or an anthracycline) in the absence and the presence of either verapamil or CyA. An increase in uptake of the dye in the presence of the inhibitor indicates an MDR phenotype. The MRP pump, which acts on Hoechst 33342 but not rhodamine-123, is not inhibited by CyA or verapamil. Sodium cyanide will inactivate both the MDR and the p-glycoprotein pump.

An anthracycline, such as daunomycin or doxorubicin, is frequently chosen for these experiments. Both these compounds are used for cancer chemo-therapy, bind to DNA, and fluoresce orange when excited by blue light. The cells to be studied need to be in suspension as single cells (see Chapter 3 for methods for preparing cells). For the preliminary experiments, it is helpful to have a pair of well-characterized cell lines, one sensitive to the drug and the other ex-hibiting MDR. *Protocol 4* describes a method using daunomycin as the test drug and CyA as the efflux blocker. The nuclei of dead cells will bind the drug and fluoresce brightly. The cells can be excluded from the analysis either by ignoring brightly fluorescent cells or by adding 10 µg/ml 7-aminoactinomycin D (7-AAD) (Molecular Probes) just before analysis and gating-out the cells fluorescing deep red.

Protocol 4

Measuring overexpression of the MDR1 pump using anthracyclines

Reagents

- Daunomycin
- Cyclosporin A
- 7-aminoactinomycin D (7-AAD) (Molecular Probes)

Protocol 4 continued

A. Preliminary experiments

1. Prepare a suspension of drug-sensitive cells at a concentration of about 10^6 cells/ml in tissue culture medium.

2. Incubate the cells at 37°C for 30 min with a series of concentrations of daunomycin in the range 0 to 30 μM.

3. Record orange fluorescence (see below). Select the lowest concentration that gives a level fluorescence well separated from the cells without drug (an increase in median fluorescence at least 3 × greater)—a typical concentration would be 3 μM.

4. Incubate cells at the selected concentration, taking samples at different time points over 1 h.

5. Record fluorescence and select the shortest time at which the increase in fluorescence has reached a plateau.

B. Experimental protocol

1. Divide a suspension of cells, at a concentration of about 10^6 cells/ml in tissue culture medium, into two tubes. To one tube add daunomycin at the concentration selected above (Part A, step 3), to the other add daunomycin plus 3 μM CyA.[a]

2. Incubate the cells at 37°C for the time selected above (Part A, step 5).

3. Add 7-AAD at 10 μg/ml.

4. Analyse, using an argon-ion laser tuned to 488 nm and recording right-angle and forward light scatter, deep red (670 nm) and orange (575 nm) fluorescence. (Use the filter sets for PerCP–PC5 and PE, respectively, see Chapter 2). Gate on the light scatter from single cells, gate-out the bright red cells and display a histogram of orange fluorescence.

5. Measure the median orange fluorescence in the absence and presence of cyclosporin A; an increase in fluorescence in the presence of CyA indicates an MDR phenotype.

[a] Alternatively, use verapamil at 100 μg/ml.

The method is illustrated in *Figure 5*.

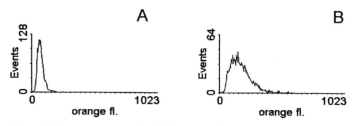

Figure 5 Functional assay for MDR using a drug-resistant variant of the human ovarian carcinoma cell line, SKOV3. The cells were incubated with 3 μg/ml daunomycin for 30 min, alone (A) or with the addition of 3 μg/ml cyclosporin A (B). (The data were supplied by Xiu-Yan Xie and David Hedley, Princess Margaret Hospital, Toronto. Analysis was on a Coulter Elite ESP; fluorescence was collected through a 575 BP filter.)

5.3 Measurement of glutathione (5)

Glutathione can be detected by its reaction with either monochlorobimane (mClB) or monobromobimane (mBrB). Both compounds are non-fluorescent and, on reaction with GSH, give a product which fluoresces blue on excitation with UV. mClB, which is the reagent of choice for rodent cells, is more specific. Its reaction with GSH is catalysed by GSH-S-transferase and it does not react with protein sulfydryls. Unfortunately, many human cells lack the appropriate isoenzyme of GSH-S-transferase and mBrB must be used. mBrB reacts directly with GSH but also reacts, albeit more slowly, with protein sulfydryls.

A method using mBrB is given in *Protocol 5* and a typical result shown in *Figure 6*.

Protocol 5

Measurement of glutathione in human cells

Reagents

- PI

- Monobromobimane (mBrB; Molecular Probes) stock solution at 4 mM in ethanol. Dilute stock solution 1:100 in water to give a working solution at 40 μM.

A. Preliminary experiment

1. Prepare a suspension of single cells at a concentration of about 10^6 cells/ml.

2. Add PI to the cells at a concentration of 5 μg/ml and mBrB (Molecular Probes) at a concentration of 40 μM.

3. Incubate at 37°C for 20 min.

4. Analyse in the flow cytometer, with a UV laser, recording right-angle and forward light scatter, blue and red fluorescence.

5. Adjust the blue PMT voltage to place the blue fluorescence in the upper half of the histogram.

6. Set a region on a cytogram of light scatter to include single cells and exclude clumps and debris. Set a region on a histogram of red (PI) fluorescence to exclude the positive cells.

7. Set gates for the histogram of blue fluorescence using the regions defined in step 6.

B. Experimental protocol

1. Prepare a suspension of single cells at a concentration of about 10^6 cells/ml. Add PI to the cells at a concentration of 5 μg/ml.

2. Set up the flow cytometer, with a UV laser, recording right-angle and forward light scatter, blue and red fluorescence, and time. Set regions on light scatter and red fluorescence as in Part A, step 6 above. Display a cytogram of blue fluorescence vs. time, gated on these regions.

Protocol 5 continued

3.   Adjust the temperature of the sample holder to 37 °C.

4.   Run the cells in the cytometer, recording and listing data, for a few seconds to establish a baseline.

5.   Leaving the cytometer recording, remove the sample and add mBrB to a concentration of 40 μM. Immediately replace the sample and continue recording.

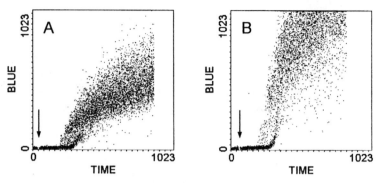

Figure 6 Measurement of glutathione in a human ovarian carcinoma cell line, A2780, using mBrB. (A) Parent line; (B) a subline resistant to the chemotherapeutic drug, cisplatin. (The cells were prepared by Jeff Holford, Institute of Cancer Research, Sutton, and the data recorded on a Coulter Elite ESP with an argon-ion ion laser tuned to give 100 mW UV.) Blue fluorescence was collected through a bandpass filter at 460 nm. The scale on the parameter TIME was 0–10 min. Data was collected for a few seconds without dye to establish a baseline; 40 μM mBrB was then added at the time marked with an arrow and data collection continued.

Acknowledgements

I thank Richard Allman and Amanda J. Taylor, Velindre Hospital, Cardiff, Wales, and Xiu-Yan Xie and David Hedley, Princess Margaret Hospital, Toronto, for supplying me with the data used to construct *Figures 4* and *5*.

References

1.   O'Hare, M. J., Ormerod, M. G., Imrie, P. R., Peacock, J. H., and Asche, A. (1989). In *Electroporation and electrofusion in cell biology* (ed. E. Neumann, A. E. Sowers, and C. Joran), p. 319. Plenum Press, New York.
2.   Robinson, J. P., Carter, W. O., and Narayanan, P. K. (1994). In *Methods in cell biology*, Vol. 42 (ed. Z. Darzynkiewicz, J. P. Robinson, and H. A. Crissman), p. 437. Academic Press, San Diego, CA.
3.   Rothe, G. and Valet, G. (1990). *J. Leucocyte Biol.*, **47**, 440.
4.   Krishan, A., Fitz, C. M., and Andritsch, I. (1997). *Cytometry*, **29**, 279.
5.   Hedley, D. and Chow, S. (1994). *Cytometry*, **15**, 349.

Appendix 1
Safety procedures

1 Bench-top flow cytometers

Bench-top cytometers are fully enclosed. The biological hazards associated with them relate to sample preparation rather than the instrument itself. Normal laboratory safety procedures should be followed when preparing and handling biological samples.

It is best, whenever the experiment permits, to fix samples before analysis. After running unfixed samples, the sample lines should be decontaminated by running a disinfectant, such as dilute bleach. The contents of the waste container should be sterilized by the addition of disinfectant before disposal in compliance with your institution's guidelines.

2 Cell sorters

The comments above on biologically hazardous materials apply. In addition, there are three hazards associated with instruments that sort cells by drop deflection (see Chapter 4)—the laser beam, the creation of aerosols during the sorting process, and the high voltage applied to the deflection plates.

2.1 Lasers

The laser beams will burn the retina of the eye and cause permanent damage. They should normally be fully enclosed. However, when they are being aligned, the beam may be exposed.

During alignment, any uninvolved personnel should be excluded from the room. No shiny metal object (which can reflect a laser beam into someone's eye) should be brought near to a laser beam; rings and bracelets should be removed or securely covered.

Safety goggles are available and should be worn whenever practicable when working with exposed beams. Local rules for the operation of lasers should be discussed and agreed with your laser safety officer.

2.2 Aerosol formation

There is a risk of aerosols containing biological particles being created from the stream emerging from the flow cell of a cell sorter. An aerosol may be created whether the instrument is actually sorting or not. There is an increased probability of aerosol formation during sorting, particularly if something goes wrong, for example, dirt in the flow chamber orifice.

While samples are being analysed or sorted, the door to the sorting chamber should be closed. Some instruments permit the application of a slight negative pressure to prevent droplets escaping into the laboratory. This should always be switched on.

The necessity for further precautions depends on the sample being analysed. These precautions should be determined in consultation with the authorized Laboratory Safety Officer.

If a sample is brought from another laboratory, any hazard likely to arise from the sample must be clearly established before commencing analysis. It is particularly important that anyone bringing transfected cells for sorting should tell the operator what genes have been introduced to the cells and the hazards associated with these genes.

2.3 High voltage

Before cleaning the area around the sorting nozzle, it is essential that the high voltage on the deflection plates is turned off. If sorting has been stopped because of a blockage, sheath fluid can be sprayed on to the plates themselves, necessitating their cleaning.

3 Chemicals

When using any chemical, normal laboratory procedures should be followed with regard to any hazard warning issued by the supplier. In particular, many procedures in this book involve the use of propidium iodide and ethidium bromide. Both these compounds are putative carcinogens and/or mutagens and should be handled appropriately, including the use of protective gloves.

Appendix 2
Flow cytometers and software

1 Commercial flow cytometers

1.1 Introduction

This section contains brief descriptions of those flow cytometers in production and being sold to a wide market. There are a few other specialist instruments available. Many laboratories run instruments that are no longer manufactured; some were produced by companies that have since withdrawn from the market; others are still supported by the manufacturer but which have been superseded by improved models.

1.2 Beckman-Coulter

This company previously traded under the name 'Coulter'.

1.2.1 Epics XL/XL-MCL

The Epics XL is a bench-top instrument equipped with a quartz-cuvette flow cell and an air-cooled argon-ion laser. The laser beam is focused and shaped using crossed-cylindrical lenses (see Chapter 1, Section 4.1). Four fluorescences can be measured. If desired, the selection of dichroic and barrier filters can be changed by the user. The conventional electronics has been replaced by transputer circuitry.

The computing platform is run on an IBM-PC compatible computer. The System II software runs under DOS; an alternative package, Expo, runs under Windows.

The MCL option has a 32-tube, carousel auto-loader, designed to work with the Coulter Multi-Q-Prep workstation for sample preparation.

1.2.2 Epics Altra

The Altra, a droplet-deflection cell sorter which can be equipped with up to four lasers, has replaced the Epics Elite. It offers a range of cuvette flow chambers with different orifice diameters, including a high-speed flow cell for faster cell sorting; 'stream-in-air' chambers of different orifice diameters are also available. The laser beams are focused and shaped using a single pair of crossed-cylindrical lenses; there are seven beam-shaping assemblies giving the user a choice of beam shape. Five fluorescences can be measured; when using more than one laser, fluorescences from two lasers can be measured on a single PMT.

The dichroic and barrier filters can be changed by the user. Pulse-processing, the ratio of two signals and time as parameters are standard.

The software (Expo) is run under Windows 95 on an IBM PC-compatible computer.

The Epics Altra Hypersort option is designed for high-speed sorting (typically at 25 000 events/sec).

1.3 Becton Dickinson

1.3.1 FACSCalibur

The FACSCalibur is a development of the bench-top instruments, the FACScan and the FACSort. It has a cuvette flow cell and a piezoelectric cell-capture system for cell sorting (see Chapter 4, Section 6) is available as an optional extra. The standard instrument has an air-cooled argon-ion laser; a second laser, air-cooled helium–neon, can be added. The laser beam is focused and shaped using an elliptical lens system, which gives a spot size of about 20 × 60 μm. In the standard configuration, three fluorescences can be measured; a fourth fluorescence detector is added if the additional laser is fitted. For automated sample handling, a 40-tube carousel (FACSloader) can be installed.

The software runs on a MAC platform.

1.3.2 FACSVantage SE

FACSVantage is a droplet-deflection cell sorter with 'stream-in-air' flow cells of different orifice diameter. It can be fitted with two lasers but, because the 'stream-in-air' configuration requires more laser power, it cannot be used with small air-cooled lasers. A third laser, necessitating an extension to the optical bench, can be installed. The laser beam is focused and shaped by an elliptical lens. Three fluorescences can be measured in the standard system. With a second laser fitted, up to three further PMTs for fluorescence measurement can be installed. The dichroic and barrier filters can be changed by the user. Pulse-processing and the ratio of two signals as an extra parameter are optional extras.

The software runs on a MAC platform.

The TurboSORT option gives high-speed sorting.

1.4 Cytomation

1.4.1 MoFlo

The MoFlo MLS was designed as a high-speed sorter, analysing 25 000 or more events/sec. It is the only instrument that can sort four populations simultaneously. It can accommodate up to three lasers, each laser has its own focusing and beam-shaping optics using its own spherical lens system so that the beams are focused on the sample stream independently (in other instruments, two or three lasers are focused through the same beam-shaping lens). It is modular in design so that extra optical components can be added; the standard power panel has space for seven PMTs. The optical filters are selectable by the user.

The software runs under Windows 95 on an IBM-PC compatible computer.

The MoFlo DTS is a lower priced model with a lower specification (slower two-way sorting).

1.5 Partec

The Partec instruments are also marketed by Dako under the name Galaxy. They use microscope-based optics with epi-illumination (see Section 2.4.3). One of the features of the Partec instruments is the low CVs obtained for DNA stains.

1.5.1 PAS

The PAS model can be used with either an air-cooled, argon-ion laser or with a UV lamp. Four fluorescences can be measured. The computing platform is based on Windows 95/98/NT running on an IBM PC-compatible machine.

1.5.2 PAS III

This instrument has a modular design so that the optical detection system can be built up to the user's requirement. It can be used with two or three air-cooled lasers and an arc lamp. A piezoelectric cell-sorting attachment (see Chapter 4, Section 6) is available. The software runs under Windows 95 on an IBM PC-compatible computer.

1.5.3 CCA-I and CCA-II

These are one (CCA-I)- or two-parameter (CCA-II), arc-lamp cytometers with a built-in PC computer.

1.5.4 PA-I and PA-II

These are one-parameter, arc-lamp (PA-I) or laser and arc-lamp (PA-II) cytometers with a built-in PC.

2 Software for data manipulation

Most listed data files are written using the FCS standard so that they can be read by any program designed to read files in this format. Apart from the software supplied by the manufacturers of commercial instruments, there are a variety of software packages available that allow data analysis off-line. The list below is probably not comprehensive but gives some of the major packages in use. Some software packages are free and can be downloaded over the Internet. The prices for the others vary considerably.

2.1 WINMDI

WINMDI is a free package written and maintained by Joseph Trotter (La Jolla, CA, USA). One version runs under Windows 3.1; the other under Windows 95. It can be downloaded from http://facs.scripps.edu/. It is a powerful program particularly useful for preparing figures, lecture slides, etc., as well as routine data analysis.

2.2 WinList

WinList is a comprehensive package for data analysis and display that runs under Windows 95, Windows NT, or on a Macintosh. It is sold by Verity.

2.3 WinFCM

WinFCM was written to replace the Hewlett Packard systems used on the older Becton Dickinson instruments, which are year 2000 incompatible (FACStar, FACStar Plus, and FACScan). It is sold by Applied Cytometry Systems.

2.4 MacLas and WinLas

MacLas and WinLas, supplied by Medical Science Associates, run on either a Macintosh Power PC, or under Windows 95 or Windows NT on a PC. They focus on supporting the automation of data analysis.

2.5 FlowJo

FlowJo is written for the Macintosh and is marketed by Tree Star. It can be purchased over the Internet.

2.5 Software for analysis of DNA histograms

There are various programs that can be used to deconvolve the DNA histogram into the component parts of the cell cycle (see Chapter 6, Section 5). These are three major commercial programmes:

- Modfit sold by Verity;
- Multicycle sold by Phoenix Flow Systems;
- FlowJo that contains cell-cycle routines as part of the package.

A free program is available over the Internet from Terry Hoy, Cardiff, Wales (http://www.uwcm.ac.uk/uwcm/hg/hoy/software.html).

Appendix 3
Suppliers

**A Core list of suppliers is available on-line at
http://www4.oup.co.uk/biochemistry/pas/supplier/**

Accurate Chemical and Scientific Co., 300 Shames Drive, Westbury, NY
11590, USA.
Web site: www.accurate-assi-leeches.com

Agar Scientific Ltd, 66a Cambridge Road, Stanstead, Essex CM24 8DA, UK.
Tel: 01279 813519

Althin Medical Ltd, Unit 25, Science Park, Milton Road, Cambridge CB4 4FW,
UK.

Amersham Pharmacia Biotech

Amersham Pharmacia Biotech, Amersham Place, Little Chalfont, Buckinghamshire
HP7 9NA, UK.
Tel: 0800 515 313   Web site: www.apbiotech.com

Amersham Pharmacia Biotech, 800 Centennial Avenue, PO Box 1327, Piscataway, NJ
08855, USA.
Web site: www.apbiotech.com

Amicon, Millipore Corporation, 80 Ashby Road, Bedford, MA 01730, USA.
Web site: www.millipore.com/analytical/amicon/index.html

Anderman and Co. Ltd, 145 London Road, Kingston-upon-Thames, Surrey
KT2 6NH, UK.
Tel: 0181 541 0035   Fax: 0181 541 0623

Applied Cytometry Systems, Dinnington Business Centre, Outgang Lane,
Dinnington, S31 7QY, UK.
Web site: www.appliedcytometry.com

Baxter Health Care Corp., Thetford, Norfolk, UK.

BDH Laboratory Supplies, Poole, Dorset BH15 1TD, UK.
Tel: +44 1202 660444   Fax: +44 1202 666856   Web site: www.bdh.com

Beckman Coulter Inc.

Beckman Coulter Inc., 4300 N. Harbor Boulevard, PO Box 3100, Fullerton, CA
92834-3100, USA.
Tel: 001 714 871 4848   Fax: 001 714 773 8283
Web site: www.beckman.com

Beckman Coulter (UK) Ltd, Oakley Court, Kingsmead Business Park, London Road,
High Wycombe, Buckinghamshire HP11 1JU, UK.
Tel: 01494 441181   Fax: 01494 447558   Web site: www.beckman.com

Becton Dickinson and Co.

Becton Dickinson and Co., 21 Between Towns Road, Cowley, Oxford OX4 3LY, UK.
Tel: 01865 748844   Fax: 01865 781627   Web site: www.bd.com

Becton Dickinson and Co., 1 Becton Drive, Franklin Lakes, NJ 07417-1883, USA.
Tel: 001 201 847 6800   Web site: www.bd.com

The Binding Site Ltd, PO Box 4073, Birmingham B29 6AT, UK.

Bio 101 Inc.

Bio 101 Inc., c/o Anachem Ltd, Anachem House, 20 Charles Street, Luton, Bedfordshire LU2 0EB, UK.
Tel: 01582 456666   Fax: 01582 391768   Web site: www.anachem.co.uk
Bio 101 Inc., PO Box 2284, La Jolla, CA 92038-2284, USA.
Tel: 001 760 598 7299   Fax: 001 760 598 0116   Web site: www.bio101.com

Bio-Ergonomics Inc., 4280 Centerville Road, St. Paul, MN 55127, USA.

Biogenex, 4600 Norris Canyon Road, San Ramon, CA 94583, USA.

Biomen Diagnostics, Pentos House, Falcon Business Park, Ivanhoe Road, Finchampstead, Berkshire RG40 4QQ, UK.

Bio-Rad Laboratories Ltd

Bio-Rad Laboratories Ltd, Bio-Rad House, Maylands Avenue, Hemel Hempstead, Hertfordshire HP2 7TD, UK.
Tel: 0181 328 2000   Fax: 0181 328 2550   Web site: www.bio-rad.com
Bio-Rad Laboratories Ltd, Division Headquarters, 1000 Alfred Noble Drive, Hercules, CA 94547, USA.
Tel: 001 510 724 7000   Fax: 001 510 741 5817   Web site: www.bio-rad.com

Bio-Whittaker UK Ltd, 1 Ashville Way, Wokingham, Berkshire RG41 2PL, UK.

Boehringer

Boehringer, Bell Lane, Lewes, East Sussex BN7 1LG, UK.
Boehringer, 9115 Hague Road, Indianapolis, IN 46250, USA.

BPL Bio Products, Dagger Lane, Elstree, Hertfordshire WD6 3BX, UK.

British BioCell International Ltd, Golden Gate, Ty Glas Avenue, Cardiff CF4 5DX, UK.
Tel: +44 (0) 1222 747232

Calbiochem

Calbiochem, PO Box 12087, San Diego, CA 92112-4180, USA.
Calbiochem, Boulevard Industrial Park, Padge Road, Beeston, Nottingham. NG9 1BR, UK.
Web site: www.calbiochem.com

Caltag Laboratories, 1849 Bayshore Boulevard #2000, Burlingame, Ca 94010, USA.
Web site: www.caltag.com

Cambridge Bioscience, 24-25 Signet Court, Newmarket Road, Cambridge CB5 8LA, UK.

CellPro, St-Pietersplein 11/12, B-1970 Wezembeek-Oppem, Belgium.

Cobe International, Blood Component Technology, Mercuriusstraat 30, 1930 Zaventum, Belgium.

Corning Inc., Science Products Division, 45 Nagog Park, Acton, MA 01720, USA.
Web site: www.corningcostar.com

CP Instrument Co. Ltd, PO Box 22, Bishop Stortford, Hertfordshire CM23 3DX, UK.

Tel: 01279 757711 Fax: 01279 755785

Web site: www.cpinstrument.co.uk

Cytek Development, Topsfield, MA, USA.

Cytomation, Fort Collins, Colorado, USA.

Web site: www.cytomation.com

Dako

Dako Ltd, Denmark House, Angel Drove, Ely, Cambridge CB7 4ET, UK.

Web site: www.dakoltd.co.uk

Dako Corp., 6392 Via Road, Carpinteria, CA 93013, USA.

Web site: www.dakousa.com

Diachem International Ltd, Unit 5, Gardiners Place, West Gillibrands, Skelmersdale, Lancashire WN8 9SP, UK.

Diatome Ltd, 2501 Bienne, PO Box 551, Switzerland.

Drukker International, Beversestraat 20, 5431 SH Cuijk, The Netherlands.

Tel: +31(0) 485 39 57 00

Dupont

Dupont (UK) Ltd, Industrial Products Division, Wedgwood Way, Stevenage, Hertfordshire SG1 4QN, UK.

Tel: 01438 734000 Fax: 01438 734382 Web site: www.dupont.com

Dupont Co. (Biotechnology Systems Division), PO Box 80024, Wilmington, DE 19880-002, USA.

Tel: 001 302 774 1000 Fax: 001 302 774 7321 Web site: www.dupont.com

Eastman Chemical Co., 100 North Eastman Road, PO Box 511, Kingsport, TN 37662-5075, USA.

Tel: 001 423 229 2000 Web site: www.eastman.com

Elga Ltd, Lane End, High Wycombe, Buckinghamshire HP14 3JH, UK.

Fisher Scientific

Fisher Scientific UK Ltd, Bishop Meadow Road, Loughborough, Leicestershire LE11 5RG, UK.

Tel: 01509 231166 Fax: 01509 231893 Web site: www.fisher.co.uk

Fisher Scientific, Fisher Research, 2761 Walnut Avenue, Tustin, CA 92780, USA.

Tel: 001 714 669 4600 Fax: 001 714 669 1613

Web site: www.fishersci.com

Fluka

Fluka, PO Box 2060, Milwaukee, WI 53201, USA.

Tel: 001 414 273 5013 Fax: 001 414 2734979

Web site: www.sigma-aldrich.com

Fluka Chemical Co. Ltd, PO Box 260, CH-9471, Buchs, Switzerland.

Tel: 0041 81 745 2828 Fax: 0041 81 756 5449

Web site: www.sigma-aldrich.com

Greiner Labortechnik Ltd, Brunel Way, Stroudwater Business Park, Stonehouse, Gloucester GL10 3SX, UK.

Tel: 01453 825255

Harlan Sera-Lab Ltd

Harlan Sera-Lab Ltd, Dodgeford Lane, Belton, Loughborough, LE12 9TE, UK.

Harlan Sprague-Dawley Inc., PO Box 29176, Indianapolis, IN 46229-0176, USA.

Web site: www.harlan.com

Hybaid

Hybaid Ltd, Action Court, Ashford Road, Ashford, Middlesex TW15 1XB, UK.

Tel: 01784 425000   Fax: 01784 248085   Web site: www.hybaid.com

Hybaid US, 8 East Forge Parkway, Franklin, MA 02038, USA.

Tel: 001 508 541 6918   Fax: 001 508 541 3041   Web site: www.hybaid.com

HyClone Laboratories, 1725 South HyClone Road, Logan, UT 84321, USA.

Tel: 001 435 753 4584   Fax: 001 435 753 4589

Web site: www.hyclone.com

Invitrogen

Invitrogen BV, PO Box 2312, 9704 CH Groningen, The Netherlands.

Tel: 00800 5345 5345   Fax: 00800 7890 7890

Web site: www.invitrogen.com

Invitrogen Corp., 1600 Faraday Avenue, Carlsbad, CA 92008, USA.

Tel: 001 760 603 7200   Fax: 001 760 603 7201

Web site: www.invitrogen.com

Kirkegaard and Perry Laboratories, Inc., (KPL) 2 Cessna Court, Gaithersburg, MD 20897, USA.

Web site: www.kpl.com

Krüess Werkstätten für Optik, Feinmechanik und Elektronik GmbH, Alster-
dorferstrasse 220, D-22297 Hamburg, Germany

Tel: 0049 040 514 31 70   Fax: 0049 040 51 25 22   Web site: www.kruess.com

Leica, Davy Avenue, Knowlhill, Milton Keynes MK5 8LB, UK.

Tel: 01908 666663   Web site: www.leica.com

Life Technologies

Life Technologies Ltd, PO Box 35, Free Fountain Drive, Incsinnan Business Park,
Paisley PA4 9RF, UK.

Tel: 0800 269210   Fax: 0800 838380   Web site: www.lifetech.com

Life Technologies Inc., 9800 Medical Center Drive, Rockville, MD 20850, USA.

Tel: 001 301 610 8000   Web site: www.lifetech.com

Lockertex, PO Box 161, Church Street, Warrington, Cheshire, WA1 2SU,
UK.

Lorne Diagnostics Ltd, Bury St Edmunds, Suffolk, UK.

Medical Science Associates, 6565 Penn Avenue, Pittsburgh, PA 15206, USA.

Tel: 001 412 362 9840.   Fax: 001 412 362 0536

Web site: www.msa.com/medsci

Merck Sharp & Dohme

Merck Sharp & Dohme Research Laboratories, Neuroscience Research Centre, Ter-
lings Park, Harlow, Essex CM20 2QR, UK.

Web site: www.msd-nrc.co.uk

MSD Sharp and Dohme GmbH, Lindenplatz 1, D-85540, Haar, Germany.

Web site: www.msd-deutschland.com

Millipore

Millipore (UK) Ltd, The Boulevard, Blackmoor Lane, Watford, Hertfordshire WD1 8YW, UK.

Tel: 01923 816375   Fax: 01923 818297

Web site: www.millipore.com/local/UK.htm

Millipore Corp., 80 Ashby Road, Bedford, MA 01730, USA.

Tel: 001 800 645 5476   Fax: 001 800 645 5439

Web site: www.millipore.com

Miltenyi BiotecGmbH, Friedrich Ebert-Str. 68, D-51429 Bergisch Gladbach, Germany

Tel: +49 2204 83060   Fax: +49 2204 85197

Web site: www.miltenyibiotec.com

Molecular Probes

Molecular Probes Inc., PO Box 22010, Eugene, OR 97402-0469, USA.

Molecular Probes Europe BV, PoortGebouw, Rijnsburgerweg 10, 2333 AA Leiden, The Netherlands.

Web site: www.probes.com

Nanoprobes Inc., 25 E Loop Road, Sye. 124, Stony Brook, NY 11790-3350 USA.

National Diagnostics

National Diagnostics, Unit 3, Chamberlain Road, Aylesbury, HP19 3DY, UK.

National Diagnostics, 10113-1017 Kennedy Boulevard, Manneville, NJ 08835, USA.

NEN™, Life Science Products, 549-3 Albany Street, Boston, MA 02118, USA.

Web site: www.nenlifesci.com

New England Biolabs, 32 Tozer Road, Beverley, MA 01915-5510, USA.

Tel: 001 978 927 5054

Nikon

Nikon Corp., Fuji Building, 2-3, 3-chome, Marunouchi, Chiyoda-ku, Tokyo 100, Japan.

Tel: 00813 3214 5311   Fax: 00813 3201 5856

Web site: www.nikon.co.jp/main/index_e.htm

Nikon Inc., 1300 Walt Whitman Road, Melville, NY 11747-3064, USA.

Tel: 001 516 547 4200   Fax: 001 516 547 0299

Web site: www.nikonusa.com

Nycomed

Nycomed Amersham plc, Amersham Place, Little Chalfont, Buckinghamshire HP7 9NA, UK.

Tel: 01494 544000   Fax: 01494 542266

Web site: www.amersham.co.uk

Nycomed Amersham, 101 Carnegie Center, Princeton, NJ 08540, USA.

Tel: 001 609 514 6000   Web site: www.amersham.co.uk

Nycomed AS Pharma, Diagnostic Division, PO Box 4284 Torshov, N-0401 Oslo, Norway.

Ortho-Clinical Diagnostics, 1001 US Highway, PO Box 250, Raritan, NJ 08869, USA.

Ortho Diagnostic Systems, PO Box 653, Enterprise House, Station Road, Loudwater, Buckinghamshire HP10 9XH, UK.

Oxoid Ltd, Basingstoke, Hampshire, UK.

Partec Gmbh, Otto-Hahn-Strasse 32, D-41861 Munster, Germany

Web site: www.partec.de

Perkin Elmer Ltd, Post Office Lane, Beaconsfield, Buckinghamshire HP9 1QA, UK.

Tel: 01494 676161   Web site: www.perkin-elmer.com

PerSeptive Biosystems Inc., 500 Old Connecticut Path, Framingham, MA 01701, USA.

Web site: www.pbio.com

Pharmacia

Pharmacia Biotech (Biochrom) Ltd, Unit 22, Cambridge Science Park, Milton Road, Cambridge CB4 0FJ, UK.

Tel: 01223 423723   Fax: 01223 420164   Web site: www.biochrom.co.uk

Pharmacia and Upjohn Ltd, Davy Avenue, Knowlhill, Milton Keynes, Buckingham-shire MK5 8PH, UK.

Tel: 01908 661101   Fax: 01908 690091   Web site: www.eu.pnu.com

Pharmacia, 23 Grosvenor Road, St Albans, Hertfordshire AL1 3AW, UK.

Pharmingen: distributed by Cambridge Bioscience

Phoenix Flow Systems, 11575 Sorrento Valley Road, #208, San Diego, CA 92121, USA (www.phnxflow.com).

Pierce Chemical Co., 3747 N. Meridian Road, Rockford, IL 61105, USA.

Web site: www.piercenet.com

Pierce and Warriner (UK) Ltd, 44 Upper Northgate Street, Chester CH1 4EF, UK.

Promega

Promega UK Ltd, Delta House, Chilworth Research Centre, Southampton SO16 7NS, UK.

Tel: 0800 378994   Fax: 0800 181037   Web site: www.promega.com

Promega Corp., 2800 Woods Hollow Road, Madison, WI 53711-5399, USA.

Tel: 001 608 274 4330   Fax: 001 608 277 2516

Web site: www.promega.com

Qiagen

Qiagen UK Ltd, Boundary Court, Gatwick Road, Crawley, West Sussex RH10 2AX, UK.

Tel: 01293 422911   Fax: 01293 422922   Web site: www.qiagen.com

Qiagen Inc., 28159 Avenue Stanford, Valencia, CA 91355, USA.

Tel: 001 800 426 8157   Fax: 001 800 718 2056

Web site: www.qiagen.com

R & D Systems, 614 McKinley Place NE, Minneapolis, MN 55413, USA.

Raymond A. Lamb, 6 Sunbeam Road, London NW10 6SL, UK.

Roche Diagnostics

Roche Diagnostics Ltd, Bell Lane, Lewes, East Sussex BN7 1LG, UK.

Tel: 01273 484644   Fax: 01273 480266   Web site: www.roche.com

Roche Diagnostics Corp., 9115 Hague Road, PO Box 50457, Indianapolis, IN 46256, USA.

Tel: 001 317 845 2358   Fax: 001 317 576 2126   Web site: www.roche.com

Roche Diagnostics GmbH, Sandhoferstrasse 116, 68305 Mannheim, Germany.

Tel: 0049 621 759 4747   Fax: 0049 621 759 4002   Web site: www.roche.com

Schleicher and Schuell Inc., Keene, NH 03431A, USA.

Tel: 001 603 357 2398

Sefar Inc, CH-8803 Rüschlikon, Switzerland

Web site: www.sefar.com

Serotec Ltd, 22 Bankside, Station Approach, Kidlington, Oxford OX5 1JE, UK.

Shandon Scientific Ltd, 93–96 Chadwick Road, Astmoor, Runcorn, Cheshire WA7 1PR, UK.

Tel: 01928 566611   Web site: www.shandon.com

Sigma-Aldrich

Sigma-Aldrich Co. Ltd, The Old Brickyard, New Road, Gillingham, Dorset XP8 4XT, UK.

Tel: 01747 822211   Fax: 01747 823779

Web site: www.sigma-aldrich.com

Sigma-Aldrich Co. Ltd, Fancy Road, Poole, Dorset BH12 4QH, UK.

Tel: 01202 722114   Fax: 01202 715460

Web site: www.sigma-aldrich.com

Sigma Chemical Co., PO Box 14508, St Louis, MO 63178, USA.

Tel: 001 314 771 5765   Fax: 001 314 771 5757

Web site: www.sigma-aldrich.com

Small Parts, PO Box 4650, Miami Lakes, FL 33014-0650, USA.

Web site: www.smallparts.com

Sorvall Centrifuges (distributors): Medi-Tech International, Inc., 2924 NW 109th Avenue, Miami, FL 33172, USA.

Web site: www.sorvall.com

Stedim, Z.I. des Paluds, BP1051-13781, Aubagne, France.

Stratagene

Stratagene Europe, Gebouw California, Hogehilweg 15, 1101 CB Amsterdam Zuidoost, The Netherlands.

Tel: 00800 9100 9100   Web site: www.stratagene.com

Stratagene Inc., 11011 North Torrey Pines Road, La Jolla, CA 92037, USA.

Tel: 001 858 535 5400   Web site: www.stratagene.com

Streck Laboratories Inc., 14124 Industrial Road, Omaha, NE 68144, USA.

TAAB Laboratories Equipment Ltd, 3 Minerva House, Calleva Park, Aldermaston, Berkshire RG7 8NA, UK.

Tel: 0118 981 7775

Therapeutic A. L. Centre, Oxford University, Old Road, Headington, Oxford OX3 7JT, UK.

Tree Star Inc., 20 Winding Way, San Carlos CA 9407, USA.

Web site: www.treestar.com/flowjo

Tropix, 47 Wiggins Avenue, Bedford, MA 01730, USA.
Web site: www.tropix.com
United States Biochemical, PO Box 22400, Cleveland, OH 44122, USA.
Tel: 001 216 464 9277
Vector
Vector, 30 Ingold Road, Burlingame, CA 94010, USA.
Vector, 16 Wulfric Square, Bretton, Peterborough PE3 8RF, UK.
Verity Software House Inc., PO Box 247, 45A Augusta Road, Topsham, ME
    04086, USA.
Web site: www.vsh.com
Wallac Inc., 9238 Gaither Road, Gaithersburg, MD 20877, USA.
Web site: www.wallac.com
Western Laboratory Service Ltd, Unit 8, Redan Hill Estate, Redan Road,
    Aldershot, Hampshire, UK.
Tel: 01252 312128

Index

Printed in the United Kingdom
by Lightning Source UK Ltd.
111000UKS00001B/37-60